Environmental Footprints and Eco-design of Products and Processes

Series editor

Subramanian Senthilkannan Muthu, SGS Hong Kong Limited, Hong Kong, Hong Kong SAR

T0205575

More information about this series at http://www.springer.com/series/13340

Miguel Angel Gardetti
Subramanian Senthilkannan Muthu
Editors

Ethnic Fashion

 Springer

Editors
Miguel Angel Gardetti
Center for Studies on Sustainable Luxury
Buenos Aires
Argentina

Subramanian Senthilkannan Muthu
Environmental Services Manager-Asia
SGS Hong Kong Limited
Hong Kong
Hong Kong

ISSN 2345-7651   ISSN 2345-766X   (electronic)
Environmental Footprints and Eco-design of Products and Processes
ISBN 978-981-10-9245-9   ISBN 978-981-10-0765-1   (eBook)
DOI 10.1007/978-981-10-0765-1

Printed on acid-free paper

This Springer imprint is published by Springer Nature
The registered company is Springer Science+Business Media Singapore Pte Ltd.

Preface

The aim of this book is to make known different aboriginal cultures and how fashion (and luxury) can become a vehicle to rescue and revalue them. In the words of the great Mexican designer Carmen Rion[1]—recognized in 2015 for her contribution to sustainable development in Latin America within the framework of the IE Award for Sustainability in the Premium and Luxury Sectors[2]—*"they are indigenous artisans."*[3] She also stated that *"the true Mexican textile designers are the aboriginal artisans of this country"* and *"the world lost the wisdom of everyday life, of femininity, of dignity—values which are deeply rooted in aboriginal groups"* (García Muñoz 2015).

Unfortunately, in many countries, instead of revaluing aboriginal cultures, local designers often travel in search of inspiration or "trends" for new designs, unaware of the fact that the adoption of aboriginal fashion styles can provide multiple benefits to fashion brands. The development of sustainable practices, the integration with a cross-cultural design, and the creation of new market niches are difficult to learn for postmodern sensitivities (Corcuera and Dasso 2008).

This book begins with the chapter by Miguel Angel Gardetti and Shams Rahman entitled "Sustainable Luxury Fashion: The Vehicle to Revalue Aboriginal Culture" which presents a number of real world case studies—Pachacuti (UK), Carmen Rion (México), Aïny (France), Loro Piana (Italy), Ermenegildo Zegna (Italy), and Hermès (France)—to demonstrate how sustainable luxury fashion can become a vehicle for salvaging and revaluing indigenous cultures.

The following chapter written by Denise Green and entitled "Fashion(s) from the Northwest Coast: Nuu-chah-nulth Design Iterations" explores the history of Nuu-chah-nulth First Nations specifically and analyzes their distinctive aesthetics and design practice through the lens of fashion theory. The chapter concludes with a discussion of contemporary Nuu-chah-nulth designers and the circulation

---

[1]Visit http://www.carmenrion.com/.

[2]Visit http://www.ie.edu/ie-luxury-awards/.

[3]Private communication between Carmen Rion and Miguel Angel Gardetti.

of their work. In the third chapter, entitled "Korean Traditional Fashion Inspires the Global Runway," Kyung Lee presents a case study of Korean aboriginal fashion and improving the growth of Korean aboriginal fashion in the global fashion industry. This research addresses how Korean aboriginal fashion can affect sustainable fashion consumption by global consumers. Specifically, the adoption of Korean traditional natural materials, dyeing techniques, and design technologies are explored as reflected in the global fashion brands' runway collections.

In the fourth chapter, "The Sustainability and the Cultural Identity of the Fashion Product," Marlena Pop examines the thesis that cultural identity sells any product, provided that the authentic intrinsic cultural value is respected, defined, and promoted, because the European cultural economy is not only a desideratum and a top strategy—it is also a dynamic multicultural reality directed towards the sustainability of heritage values.

In the fifth chapter "Badaga Ethnic (Aboriginal) Fashion as a Local Strength," H. Gurumallesh Prabu and G. Poorani showcase the unique badaga system with special reference to attire that has been followed over centuries as fashion. Irrespective of age, all badagas tend to wear traditional attire in the form of thundu, mundu, dhuppatti, seelai, mandarai, mandepattu, etc., on almost all social occasions and perform traditional badaga dances which represent a local strength as well as ethnic fashion.

In the sixth chapter, "Ethnic Styles and Their Local Strengths," Nithyaprakash V. and Thilak Vadicherla make an in-depth review of ethnic dress over the world, analyzing its meaning.

In the seventh and final chapter, "Continuing *Geringsing* Double *Ikat* Production in Tenganan, Bali," Kaye Crippen documents the continual changes in weaving in the village of Tenganan Pegeringsingan in Bali, Indonesia from 1985 to 1999 and in 2014. The village has a centuries old tradition of producing difficult to make *geringsing* double *ikat* textiles which requires tie-dyeing of both warp and weft yarns in both the warp and weft directions to create a pattern. Reasons for the decline and subsequent partial revival of weaving are explored.

References

Corcuera R, Dasso MC (2008) Introducción. In: Corcuera R, Dasso MC (eds) Tramas Criollas, Ediciones CIAFIC, Buenos Aires, pp 9–22
García Muñoz I (2015) Las verdaderas diseñadoras textiles mexicanas, son las artesanas indígenas de este país – Interview with Carmen Rion on Caras de la Información. Source: http://carasdelainformacion.com/2015/02/10/carmen-rion-las-verdaderas-disenadoras-textiles-mexicanas-son-las-artesanas-indigenas-de-este-pais/. Accessed 8 Dec 2015

Contents

Sustainable Luxury Fashion: A Vehicle for Salvaging and Revaluing Indigenous Culture

Miguel Angel Gardetti and Shams Rahman

Abstract Sustainable luxury is coming back into favor, essentially with its ancestral meaning, i.e., thoughtful purchasing, with consideration of artisan style manufacturing, assessment of product beauty in its broadest sense, and respect for social and environmental issues. In addition, it also means consideration of craftsmanship and innovation of different nationalities and preservation of local and ancestral cultural heritage. The relationship between luxury, textiles, and fashion is quite an ambiguous one, as textiles and fashion do not fully belong to the luxury world but overlap with luxury in its most expensive and exclusive segments. Both luxury and fashion share the common need for social differentiation, but they also differ in two major aspects. First, luxury is timeless whereas fashion is ephemeral. Second, luxury is for self-reward whereas fashion is not. Thus, the term 'luxury-fashion' seems to consist of two inherently contradictory expressions, i.e., as a luxury product it is supposed to last, although as a fashion product it is expected to change frequently. Nevertheless, because the essence of fashion is change, luxury fashion gives exclusive access to enforced change. Luxury fashion is recurrent change at its highest level, and it is distinguished from other luxury segments by its constant pressure for change. However, beyond these contradictions, luxury fashion should not necessarily come into conflict with sustainable principles. In this chapter we present a number of real-world case studies—Pachacuti (UK), Carmen Rion (México), Aïny (France), Loro Piana (Italy), Ermenegildo Zegna (Italy), and Hermès (France)—to demonstrate how sustainable luxury fashion can become a vehicle for salvaging and revaluing indigenous cultures.

M.A. Gardetti (✉)
Center for Studies on Sustainable Luxury, Av. San Isidro 4166, PB "A",
C1429ADP Buenos Aires, Republic of Argentina
e-mail: mag@lujosustentable.org
URL: http://www.lujosustentable.org

S. Rahman
School of Business IT and Logistics, College of Business, RMIT University,
Melbourne, Australia
e-mail: shams.rahman@rmit.edu.au

© Springer Science+Business Media Singapore 2016
M.A. Gardetti and S.S. Muthu (eds.), *Ethnic Fashion*, Environmental Footprints
and Eco-design of Products and Processes, DOI 10.1007/978-981-10-0765-1_1

Keywords Indigenous culture · Luxury · Luxury fashion · Sustainable luxury

1 Introduction

Sustainable development is a new paradigm, and this requires looking at things from a different perspective. Although luxury has always been important as a social determinant, it is currently starting to give the opportunity to people to express their innate values. Thus, sustainable luxury promotes a return to the essence of luxury with its ancestral meaning, i.e., a thoughtful purchase, artisan manufacturing, beauty of materials in its broadest sense, and the respect for social and environmental issues. So, sustainable luxury would not only be the vehicle for greater respect for the environment and social development, but also a synonym for culture, art, and innovation of different nationalities, maintaining the legacy of local craftsmanship (Gardetti 2011). Likewise, sustainable fashion is an approach to the fashion system intended to minimize adverse social and environmental impacts along the value chain.

This chapter is organized as follows. First, we provide generic definitions of luxury and sustainable luxury and introduce readers to the world of fashion and indigenous culture. Following this, we present a number of real-world case studies describing how international luxury brands such as the established Goliaths have incorporated ancestral cultures into their products. We then provide an overview of how indigenous cultures are being salvaged and revalued by social entrepreneurs such as the emerging Davids. Finally, we analyze how sustainable luxury fashion can become a vehicle for salvaging and revaluing indigenous cultures.

2 Methodology

To develop this chapter, the authors have used academic literature as well as qualitative and quantitative information about real-world cases. Qualitative and quantitative information was collected from three different sources—corporate documents, information publicly available on the Internet, and representatives of the companies who participated in the IE Award for Sustainability in the Premium and Luxury Sectors (formed Best Performance in Sustainable Luxury in Latin America Award).[1]

---

[1]The criteria needed to be met to receive the Award are:
–*Social aspects*: the strategies carried out by the company underscoring positive impacts
–*Environmental aspects*: the strategies carried out by the company underscoring positive impacts
–*Economic aspects*: upfront investment; sales volume, profits (as a percentage of revenues), future growth expectations based on company performance, average price of product/s that your company sells, and distribution or sales channel

3 Luxury, Luxury Fashion, and Sustainable Development

The perception of luxury depends on cultural, economic, and geographical contexts. This makes luxury an ambiguous and abstract notion (Low undated; Scheibel undated). Thus, luxury is a matter of seeing and being seen. 'Seeing' can be regarded as the latest distinctive signifier for use to define 'be seen' in different distinctive group practices (Mortelmans 2005). Berry (1994) in his work *The Idea of Luxury*—one of the most comprehensive works on the concept and intellectual history of luxury—establishes that luxury is the reflection of social norms and aspirations and thus has changed in meaning over time. True elements (authentic) of luxury rely on the search for beauty, refinement, innovation, purity, and the well-made (Girón 2012). However this notion of luxury has given way to the new luxury through democratization or massification that occurred when family and artisanal luxury companies failed to compete against the large conglomerates with strong economic focus (Gardetti and Muthu 2015). *It means that the image—neither reputation nor legitimacy—was the way, and marketing was the function* (Gardetti and Torres 2014). Rahman and Yadlapalli (2015, p. 188), along the same lines as Gardetti and Torres (2014), suggest that the '*luxury industry relies heavily on communications for branding and marketing.*'

Giacosa (2014), however, reminds us that in order to set the context of luxury fashion and sustainable luxury fashion, it is necessary to differentiate the terms fashion and luxury. According to Fletcher (2008, 2014), fashion is the way in which our clothes reflect and communicate our individual vision within a society, linking us to time and space. Clothing is the material thing giving fashion a contextual vision in a society (Cataldi et al. 2010). Luxury represents items perceived as a symbol of status. It is also a symbol of elegance and sophistication with the emphasis on the intrinsic value of many categories.

The relationship between luxury and fashion is quite an ambiguous one, as fashion does not fully belong to the luxury world but overlaps with luxury in its most expensive and exclusive segments (Godart and Seong 2014). According to Kapferer (2012), both luxury and fashion share the common need for social differentiation, but they differ in two major respects: first, luxury is timeless, fashion is ephemeral, and second, luxury is for self-reward, fashion is not. Thus, luxury fashion seems to be a contradictory concept. As luxury, it is supposed to last, although as fashion it is supposed to change frequently. However, because the essence of fashion is change, luxury fashion grants exclusive access to enforced change. Luxury fashion is recurrent change at its highest level, and hence is distinguished from other luxury segments by its constant pressure for change. This idea is shared by Pinkhasov and Nair (2014) who stress that, in a celebrity-driven culture, fashion has come to dominate the image and attitude of luxury.

Sustainable development is a complex manifestation and it is hard to come to a consensus on its meaning. Each individual can consider the term and 'reinvent' it, reflecting upon his/her own needs. The ambiguous nature of the term has led us to change our objectives and priorities constantly over time. Nevertheless, one of the most widely accepted definitions of sustainable development is the one proposed by the World Commission. In its report on Environment and Development—called *Our Common Future*—defines sustainable development as the development model that allows us *to meet the present needs, without compromising the ability of future generations to meet their own needs*. The essential objective of this development model is to raise the quality of life by long-term maximization of the productive potential of ecosystems with the appropriate and relevant technologies (Gardetti 2005).

Authors such as Walker (2006) and Koefoed and Skov (undated) have studied the contradictions between fashion and sustainability and suggest that fashion should not necessarily come into conflict with sustainable principles. Indeed, fashion plays a role in the promotion and achievement of sustainability, and may even be key to more sustainable living. According to Hethorn and Ulasewicz (2008), fashion is a process which is expressed as a material object with a direct link to the environment. It is embedded in everyday life. Therefore, sustainability in fashion means that the development and use of a thing or a process are not harmful to people or the planet, and once put into action, such a thing or process can rather enhance the well-being of those who interact with it.

In the literature, it is argued that luxury and sustainability are incompatible terms, meaning both cannot be achieved at the same time. This argument is based on the fact that consumption of anything more than basic needs may jeopardize the life of following generations and is regarded as unsustainable. Luxury products are considered a waste of resources for the pleasure of few and symbolize social inequality (Rahman and Yadlapalli 2015). However, according to Godart and Seong (2014), luxury can offer a unique opportunity for creating sustainable business environments because of its two core features that set it apart from other market segments or industries. First, luxury is (often) based on unique skills. This allows luxury to provide high quality and rewarding business conditions. Second, luxury is characterized by its unique relationship with time, for its value is over the long term. This allows luxury to offer a sustainable business model for resource management and high quality product development which are two pertinent elements of sustainable luxury. Kleanthous (2011) highlights the fact that luxury has become less exclusive and less wasteful, and more about helping people express their innate values. Sustainable luxury is thus the return to the ancestral essence of luxury, i.e., a thoughtful purchase, artisan manufacture, beauty of materials in its broadest sense, and respect for social and environmental issues.

4 Overview of Fashion and Culture

Fashion is profoundly a social experience which invites individual and collective bodies to assume certain identities and, at times, also to transgress limits and create new ones (Root 2005). On the other hand, a textile is an agent of collective, gestural, and symbolic expression (Hughes 1996) which can be perceived as a language that weaves notions and concepts (Cutuli 2008). Moreover, craftsmanship is a social construct representing the cultural heritage of every region. Its expression, communication, and trade require specific channels, and demand cultural protection for both the artisans and consumers of these products in the global context. Handicrafts imply the recognition and respect for one's own local characteristics and for the typical products that express and keep alive the culture of every region of the world (Gardetti 2015).

Rapid modernization, tourism, and globalization have altered the ways in which artisans create, consume, and market traditional ritual clothes and "ethnic" dress (Root 2005). Several investigators such as Popelka and Littrell (1991) and Swain (1993) have revealed that such alteration erodes traditional cultures. Nonetheless, most artisan groups, including indigenous artisans, wish to preserve the deep-rooted local values as well as their beliefs in their relationships with the society and the environment. In the global marketplace, pieces of handicraft are being purchased by consumers who share those values, reject large-scale manufacture and mass production, and look for authentic 'local' handmade objects (Grimes and Milgram 2000). According to Guldager (2015, p. 92) *'The small differences and dissimilarities resulting from garments that are personalized or handmade are little treasures in an ocean of homogenized products. Garments that are fully handcrafted or have a craft integrated into their design might indeed be defined as a 'cultural luxury.' ... Besides the essential and rare beauty in crafted garments and the emotional value associated with them, the cultural process of maintaining the great heritage based on traditions which have been passed on for generations is of great value. The different ancient techniques and knowledge within these techniques and how they have been demonstrated and transferred to new generations should also be considered. This precious transference is invaluable in the process of obtaining these human-based 'cultural fortunes' that can be defined as articles of luxury'*. Although for Corcuera and Dasso (2008) this knowledge is a quality difficult to learn for postmodern sensitivities, there are many entrepreneurs and established brands trying to rescue and revalue the way indigenous ethnic groups dress.[2,3]

---

[2]There are also institutional initiatives such as, for instance, the **Ethical Fashion Initiative**. Within the International Trade Centre, the Poor Communities Trade Programme (PCTP) aims to reduce global poverty by involving micro-entrepreneurs in the developing world with international and regional trade. The Ethical Fashion Initiative is its operational arm.

The Ethical Fashion Initiative is not a charity. It facilitates dignified work at a fair wage. It does so by connecting some of the world's most marginalized artisans in Africa and Haiti with the fashion industry's top talents for mutual benefit. It also works with upcoming designers in West Africa to promote local talent and increase export capacities of the region.

Please visit: http://www.intracen.org/itc/projects/ethical-fashion/the-initiative/.

[3]Some other initiatives were conceived at the very heart of indigenous communities, such as, for example, the **Indigenous Runway Project** founded by Tina Waru. There was a growing need to empower indigenous young people with confidence, motivation, and pride so that they can embrace their hidden beauty and talent and explore career pathways in various areas of fashion, modeling, fashion design, performing arts, production, hair and makeup and styling. To date, the Indigenous Runway Project has reached other global indigenous communities, such as New Zealand, Arizona, Canada, and Africa.

Please visit: http://indigenousrunwayproject.com/.

5 Emerging Davids and Established Goliaths[4]—Incorporation of Indigenous Cultures in Sustainable Luxury Fashion

Within the luxury industry, it can be observed that a new category of firms such as the emerging Davids are attractive to consumers because they are based on new value propositions. These firms can have a big business impact because of their potential for realizing a larger market share (Villiger et al. 2000). To achieve a profound social change, the role of personal values is very important. Studies suggest that idealistic values regarding environmental and commitment to social goals can be translated into economic assets (Dixon and Clifford 2007). Generally, these firms are driven by the transformational leadership behavior which inspires and guides the fundamental transformation that sustainability requires. A dilemma in sustainability is that the management systems and design principles that we have used to organize institutions are not aligned with the underlying laws of nature as well as human nature. For this reason, it is of utmost importance to have people with a profound respect of environmental and social issues and who are well motivated to "break" the rules and promote disruptive solutions to these issues. Marshal et al. (2011, p. 6) have emphasized that *"we see such leadership as necessarily going beyond conventional notions, because it needs to be able to step outside and challenge current formulations of society and business, and because sufficiently robust change means questioning the ground we stand on."*

Pachacuti[5]—a brand created by Carry Sommers[6] in 1992, which means 'the world upside down' (in the Quechua language)—is a pioneer in ethical luxury fashion, *'providing a role model to challenge compromise and mediocrity within*

---

[4]The expression 'Emerging Davids and Established Goliaths' was adapted from the *Emerging Davids versus Greening Goliaths* work developed by Hockerts and Wüstenhagen in (2009). The authors explained the interplay between Davids and Goliaths to drive industry towards sustainable development. Metaphorically these terms refer to the two different types of organizations with respect to size, age, and objective function (Rahm and Yadlapali 2015).

[5]This part of the chapter is based both on Sommers (2014) and the correspondence between Pachacuti and the organizers of the IE Award for Sustainability in the Premium and Luxury Sectors.

[6]Following the collapse of the Rana Plaza Building in Bangladesh, and given the increasing mortality rate in such a catastrophe (April, 2014), many stories were published urging consumers to support ethical fashion as a way to improve working conditions throughout the entire supply chain. A few days later, Carry Sommers created the Fashion Revolution Day to commemorate the disaster anniversary and, since then, it has become a global movement that takes place all over the world, mobilizing the entire supply chain from cotton producers and textile workers to brands and consumers. Led by brands, retailers, activists, the press, and academics from both inside and outside the sector, good practices are celebrated, thus raising awareness about the "true" cost of fashion. In December 2013, Carry Sommers was granted the Outstanding Contribution to Sustainable Fashion Award at the House of Lords in recognition of her work both at Pachacuti and Fashion Revolution Day.

Fig. 1 Carry Sommers and the weavers. *Source* Pachacuti; published with Carry Sommer's authorization

the industry.' Pachacuti demonstrates that authentic luxury is capable of incorporating both social and environmental obligations without compromising on style. However, Pachacuti is an exception within the industry. It is a company which adheres to the highest fair trade and environmental standards and yet its products are being sold in some of the most luxurious stores around the world (Fig. 1). The brand is conceived as a consequence of its founder's trips to Ecuador to research into textile production and sustainability in traditional skills and techniques since pre-Columbian times.

Pachacuti is the first fair trade business specializing in hats. Because one of Pachacuti's main objectives from the beginning has been to promote cultural heritage through the preservation of traditional skills, Panama hats were the flagship of the brand (Fig. 2a, b).

The story of the classic Panama hat is associated with a history of century long exploitation. For this reason, hat weaving had slowly been disappearing. However, the greater demand and premium prices paid for hats have resulted in a closer relationship between the company and the indigenous knitting association. It is important to note that Pachacuti's Panama hats are made of organic *Carludovica palmata* from a community-owned plantation which encourages plant and animal biodiversity (Fig. 3).

Pachacuti offers better living conditions which help aborigines to stay with their communities and families and engage themselves in hat weaving throughout the entire agricultural cycle (Fig. 4). This is a significant shift for communities

(a)   (b)

Fig. 2   a, b Weaving hands. *Source* Pachacuti; published with Carry Sommer's authorization

Fig. 3 Removing the chlorophyll from *Carludovica palmata*. *Source* Pachacuti; published with Carry Sommer's authorization

Fig. 4 "Hice tu sombrero" ("I made your hat"). *Source* Pachacuti; published with Carry Sommer's authorization

where 60 % of children have at least one of their parents working abroad which had led to family breakdown, high rates of alcoholism, youth suicide, teenage pregnancy, and poor academic performance.

Similarly, in a recent interview (February 2015) for Caras de la Información, *Carmen Rion*, Mexican designer, founder of a brand named after herself, stated that "*the true Mexican textile designers are the indigenous artisans of this country*"... and "*that the world lost the wisdom of everyday life, of dignity—values which are deeply rooted in indigenous groups*" (Fig. 5).

For Carmen Rion it all started in 2004 when she was invited by the *Fondo Nacional para el Fomento de las Artesanías* (FONART, Mexico) to train a group of artisans in Chiapas. Since then, she has included Chiapas traditional textiles in her collections which have been recognized in fashion capitals such as Paris and London (Fig. 6).

However, this is not only about collections for Carmen Rion. She developed projects to support and salvage traditional Mexican culture. For example, 'Paisaje Mocheval' was an initiative taken by Carmen Rion that encouraged Chiapas indigenous artisans to draw the landscape of their place of origin to use it in *mochevales*, a traditional garment of the region. This has resulted in a 120-garment

Fig. 5 First canvas knitted in backstrap looms for Carmen Rion. Gómez Pérez Family 2005. *Source* Carmen Rion; published with Carmen Rion's authorization

Fig. 6 Designs for Hong Kong textile gallery. Renaissance of fashion exhibition, 2015. 2016 Spring collection. *Source* Carmen Rion; published with Carmen Rion's authorization

collection being exhibited at Museo Franz Mayer in 2011 and subsequently in many cities around the world (ARCA 2015) (Figs. 7 and 8).

After many years of traveling to and from Chiapas to deliver training and conduct workshops, Carmen Rion recently selected 40 women who previously

Fig. 7 Bargoin Museum, Clermont Ferrand, France 2014. Mocheval Landscape collection, Sna Maruch + Carmen Rion group. Some of the artisans are: Catalina Perez Hernández, Magadalena Gómez Perez, Juana Albertina López, Josefa Gómez Perez, Rosa Gómez Perez, María Gómez Hernandez, Victoria Pascuala Gómez Pérez, Estela Sánchez Hernández, and Rosa Mercedes Gómez Pérez. Collectors: Claudia Muñoz, Adriana Aguerrebere, Monica Bucio, Maria Luisa Sabau, and Rosy Laura Hernández. *Source* Carmen Rion; published with Carmen Rion's authorization

Fig. 8 Indigenous artisans in Mocheval Landscape. *Source* Carmen Rion; published with Carmen Rion's authorization

worked for 'Paisaje Mocheval' to create a group called Sna Maruch. This initiative has empowered the indigenous women to strengthen further the collaborative relationships for the current collection known as 'Carmen Rion + Sna Maruch' (GuadiaSustentable.com 2015) (Fig. 9).

Aïny Savoirs Des Peuple, which was created in France by a young entrepreneur Daniel Joutard, produces cosmetics developed from a combination of sacred plants used by the "Ashaninka" community in Peru and "Quechua" and "Achuars" communities in Ecuador. This is a unique instance of blending the local cultural ingredients with Western science. The products of this firm are organic and not tested on animals, being marketed in Europe as luxury products. The firm not only promotes sustainable development but also hold collaborative agreements with indigenous communities to engage them in the production as well as in the manufacture of packaging (Gardetti and Torres 2013a).

In turn, some international brands, *Established Goliaths*, have a proactive attitude towards the challenge of sustainability. It is observed that, in general, the industry reacts to what the market and consumers are demanding (Gardetti and Girón 2014).

Ranfgani and Guercini (2015, p. 55) offered a detailed explanation of the existing relationship between *Loro Piana*, taken over by LVMH in 2013, and ancient communities: "*The natural environments where Loro Piana finds its finest raw material are unspoiled worlds that supply inimitable resources. The Loro*

Fig. 9 Mocheval Landscape, Franz Mayer Museum, Mexico, February 2011. 2011 summer collection. Forty 100 % wool-Mochevales knitted in backstrap looms by Sna Maruch Group, rescue of ancestral techniques and hand embroidery. One hundred and fifty unique pieces of this collection have been sold. This project was presented at the University of Newcastle, Bargoin Museum, France, Metropolitan Museum, Manila, Philippines. Photos by Mauricio Jimenez,/Daniel Cruz Rion. *Source* Carmen Rion; published with Carmen Rion's authorization

Piana family, in fact, invests economic and human resources in order to identify and, then, to preserve them. This is because they are convinced that it is precisely thanks to them that it is possible to combine quality and timeless elegance, and to produce fabrics and garments destined to last more than a lifetime. A quality-based philosophy seems to inform the search for yet unexplored natural resources. Their preservation at the base of Loro Piana's sustainable orientation is thus coherent with an existing company way of being. All this explains the recent environmental projects that the company has decided to undertake. One of these concerns the opening of a subsidiary in Ulan Bator (Mongolia) to work together with nomadic tribesmen in the breeding of goats whose fleece is used in the production of cashmere clothing collections. In particular, the aim of this green field investment is to transfer to the local community of breeders all the techniques of animal husbandry in order to preserve a local ecosystem and, at the same time, to pursue the high quality of the emerging raw material. The survival of this latter depends on the conservation of the underlying rare resources rooted in naturalistic environments. The products Loro Piana makes by using Mongolia fleece are a source of pride; particularly cherished, and appreciated by consumers, are the products labeled as Loro Piana Baby Cashmere. Their fiber comes from Hircus goat kids that are between three and twelve months old."

Loro Piana also organized a consortium including Condortips, a textile producer from Arequipa (southern Peru), Lanerie Agnona SpA, an Italian knit fabric producer (currently under Ermenegildo Zegna's control), and the Government of Peru (Gardetti 2011). This consortium also committed to fund an association of breeders to achieve the following two objectives:

1. To improve production by keeping the native indigenous communities to their own regional settlement areas
2. To educate the native indigenous communities on conservation techniques and methods to conserve vicuna which has been declared an endangered species.

The project had a profound impact on the population of vicuna which has increased from 6000 in 1974 to a current level of 180,000 animals. It is expected to reach 1 million over the next 10 years, depending on the fiber demand. As a result, the income of the local communities has increased fourfold since the start of the project. In addition, the local communities have also developed new capabilities and increased efficiency (Gardetti 2011).

The results of this project have motivated the global companies to engage in other similar projects. For example, in 2009 *Ermenegildo Zegna* developed a project with 200 indigenous families from the Picotani community. This project helped the communities to develop irrigation systems to improve watering capacity in the fields where vicunas live, as well as in lagoons.

The French fashion house *Hermès* included colorful Mexican handcrafted embroideries in its silk scarves. By making use of the traditional embroideries called Tenangos from the Otomi region of Tenango de Doria in Hidalgo (Mexico), Hermes has given new perspectives to their products. These embroideries depict the daily life, rites, and ceremonies of the communities inspired by the

Fig. 10 Embroidery design "Tenangos." © *Camino a Tenango*, Gimena Romero. Thule Ediciones (2015). Published with permission

pre-Hispanic Otomi culture. The cultural essence of these communities is to live and enjoy life together amidst nature (Figs. 10, 11 and 12).

The *Museo de Arte Popular* (MAP) (Popular Art Museum) contacted craftsman Vicente Ezequiel, the only indigenous artisan who still masters the design technique of Tenango embroideries, and embroiderer Elia Tolentino, who agreed to the project so as to help their community. According to the executive, *"their biggest dream is to 'make improvements to the school San Pablo el Grande and create better living conditions for the community"* (Comunidad Textil 2011). Mexican artisans survive in poverty conditions, and Tenango de Doria is no exception. In 2011, the Otomi or Ñahñu team, as referred to in their language, traveled to Mexico City and attended the design presentation and promoted silk scarves at a global level. During the ceremony, artisans showed their outstanding design and embroidery skills and their pride in displaying an aspect of their ancient culture, customs, and traditions.

The design—called "Din tini yä zuë," the Otomi words for "the encounter between man and nature"—is made up of two kinds of embroidery, best known as Tenango, in honor of the town, and was presented in nine colors (Comunidad Textil 2011). According to the Museo de Arte Popular's spokesperson (Comunidad Textil 2011), in those days *"patterns—showcasing a wide variety of colors—were inspired by the flora and fauna of their land, as well as by rituals related to their interaction with nature, such as harvest, sowing, or prayers for rain. Some other details show celebratory scenes, like weddings or carnivals."*

Fig. 11 Embroidery design "Tenangos." © *Camino a Tenango*, Gimena Romero. Thule Ediciones (2015). Published with permission

Fig. 12 Embroidery design "Tenangos." © *Camino a Tenango*, Gimena Romero. Thule Ediciones (2015). Published with permission

Hermès regards this project as "a tribute" to the inhabitants of the Tenango town. Because of its economic and social success, Hermes, through its foundation, is seeking new projects to use the craftsmanship of other Mexican regions and even the indigenous communities of other countries. As Iveth Lagos of the company highlighted, "*We share an interest in both preserving and passing on the savoir-faire of the hands that make the designs. We share an interest in preserving and passing from one generation on to the next ancient techniques that result in excellent products*" (Comunidad Textil 2011).

6 Conclusions

Sustainability in luxury can be viewed from a wide range of perspectives, and as a result it may generate different approaches to strategy formulation (Gardetti and Torres 2013b). Furthermore, sustainability has a cultural dimension in addition to environmental, economic, and social dimensions (Dresner 2002). According to Na and Lamblin's 'Sustainable Luxury: Sustainable Crafts in a Redefined Concept of Luxury from Contextual Approach to Case Study' from 2012, "*the cultural and social dimensions reflect the sustaining elements that keep the values, traditions, and social exchanges of craft alive. The cultural sustainability of craft is about maintaining the traditional skills employed, while also demonstrating responsiveness to the everyday uses of crafts in our ordinary lives. With a synthesis of forward-thinking vision and tradition-sustaining elements, the culture surrounding craft and the culture expressed through craft can survive our increasingly mass-produced age.*" This aligns with the view of Rahman and Yadlapalli (2015), who argue that globalization along with emerging new riches has accelerated the growth in the number of luxury fashion consumers.

Artisanal practices seem to differentiate between authentic luxury and mass luxury as customers acknowledge the superiority of a handmade object and value it accordingly. Therefore, sustainable luxury appears as a connective environment, where needs, values, and cultures are collectively shared. The examples offered in this chapter transfer to them the traditional sustainable production techniques or contribute to the preservation of local ecosystems and underlying traditions, obtaining in return prestigious primary resources. Although they use a wide range of inherent skills and local techniques, they provide a bridge to emotions and feelings through encoded values and aesthetics. These are the building blocks of the brand's DNA that cannot be duplicated. Crafts also form the best argument for sustainability. They are about generating humbleness and respect for the processes of making such products, and furthermore for the human beings who have created them.

Sustainable luxury would not only be the vehicle for more respect for environment and social development but also be synonymous with culture, art, and innovation of different nationalities, maintaining the legacy of local craftsmanship (Gardetti 2011). This aligns with the inherent features of craftsmanship: *traditionalism, popular authenticity, manual prominence, individual domestic production, creative sense, aesthetic sense, and specific geographic location*

(FIDA, PRODERNOA and FLACSO 2005, p. 17). All of this shows a clear relationship between craftsmanship, indigenous culture, and sustainable luxury. Ultimately, the outcome of this relationship is a set of material goods to create and endow the world with bonds of reciprocity (Corcuera and Dasso 2008, p. 17).

References

ARCA (2015) Los Diseños de Carmen Rion Rescatan el Folclor Mexicano. http://www.arca-lab.com/conoce-el-trabajo-de-la-disenadora-mexicana-carmen-rion/. Accessed 2 Nov 2015

Berry CJ (1994) The idea of luxury—a conceptual and historical investigation. Cambridge University Press, New York

Caras de la Información (2015) Las verdaderas diseñadoras textiles mexicanas, son las artesanas indígenas de este país. http://carasdelainformacion.com/2015/02/10/carmen-rion-las-verdaderas-disenadoras-textiles-mexicanas-son-las-artesanas-indigenas-de-este-pais/. Accessed 2 Nov 2015

Cataldi C, Dickson M, Grover C (2010) Slow fashion: tailoring a strategic approach towards sustainability. Thesis submitted for completion of Master's in strategic leadership towards sustainability, School of Engineering, Blekinge Institute of Technology, Karlskrona, Sweden

Comunidad Textil (2011) Los bordados mexicanos, un hallazgo que sedujo a Hermès. http://www.comunidadtextil.com/news1/2011/03/los-bordados-mexicanos-un-hallazgo-que-sedujo-a-hermes/. Accessed 3 Feb 2014

Corcuera R, Dasso MC (2008) Introducción. In: Corcuera R, Dasso MC (eds) Tramas criollas. Ediciones CIAFIC, Buenos Aires, pp 9–22

Cutuli G (2008) El textil en el contexto cultural. In: Corcuera R, Dasso MC (eds) Tramas criollas. Ediciones CIAFIC, Buenos Aires, pp 25–34

Dixon SEA, Clifford A (2007) Ecopreneurship: a new approach to managing the triple bottom line. J Organ Change Manage 20(3):326–345

Dresner S (2002) The principles of sustainability. Earthscan, London

FIDA, PRODERNOA FLACSO (2005) El Sector de Artesanías en las Provincias del Noroeste Argentino. PRODERNOA, Buenos Aires

Flechter K (2008) Sustainable fashion and textiles—design journeys. Earthscan, London

Flechter K (2014) Sustainable fashion and textiles—design journeys, 2nd edn. Earthscan, London

Gardetti MA (2005) Sustainable development, sustainability and corporate sustainability. In: Gardetti MA (ed) Texts in corporate sustainability—integrating social, environmental and economic considerations with short and long term. LA-BELL, Buenos Aires

Gardetti MA (2011) Sustainable luxury in Latin America. Lecture delivered at the seminar sustainable luxury & design within the framework of the MBA of IE. Instituto de Empresa—Business School, Madrid

Gardetti MA (2015) Loewe: luxury and sustainable management. In: Gardetti MA, Muthu SS (eds) Handbook of sustainable luxury textiles and fashion, vol 2. Springer, Singapore, pp 1–16

Gardetti MA, Girón ME (2014) Sustainable luxury and social entrepreneurship: stories from the pioneers. Greenleaf Publishing, Sheffield

Gardetti MA, Muthu SS (2015) Introduction. In: Gardetti MA, Muthu SS (eds) Handbook of sustainable luxury textiles and fashion, vol 1. Springer, Singapore, pp vii–xi

Gardetti MA, Torres AL (2013a) Entrepreneurship, innovation and luxury: the Aïny Savoirs Des Peuple case. J Corp Citizsh Issue 52:55–75

Gardetti MA, Torres AL (2013b) Introduction. J Corp Citizsh Issue 52:55–75

Gardetti MA, Torres AL (2014) Introduction. In: Gardetti MA, Torres AL (eds) Sustainability luxury: managing social and environmental performance in iconic brands. Greenleaf Publishing, Sheffield, p 1

Giacosa E (2014) Innovation in luxury fashion family business: processes and products innovation as a means of growth. Palgrave Macmillan, New York

Girón ME (2012) Diccionario LID sobre Lujo y Responsabilidad. Editorial LID, Madrid

Godart F, Seong S (2014) Is sustainable fashion possible? In: Gardetti MA, Torres AL (eds) Sustainability luxury: managing social and environmental performance in iconic brands. Greenleaf Publishing, Sheffield, pp 12–27

Grimes KM, Milgram BL (2000) Introduction: facing the challenges of artisan production in the global market. In: Grimes KM, Milgram BL (eds) Artisans and cooperatives—developing alternate trade for the global economy. The University of Arizona Press, Tucson, pp 3–10

GuadiaSustentable.com. Carmen Rion: Una Fusión Auténticamente Mexicana. http://www.vang uardiasustentable.com/carmen-rion-una-fusion-autenticamente-mexicana/. Accessed 2 Nov 2015

Guldager S (2015) Irreplaceable luxury garments. In: Gardetti MA, Muthu SS (eds) Handbook of sustainable luxury textiles and fashion, vol 2. Springer, Singapore, pp 73–97

Hethorn J, Ulasewicz C (eds) (2008) Sustainable fashion, why now?—a conversation about issues, practices, and possibilities. Fairchild Books Inc., New York

Hockerts K, Wüstenhagen R (2009) Emerging Davids versus greening Goliaths. CBS Center for Corporate Social Responsibility (Copenhagen Business School), Frederiksberg

Hughes P (1996) Tissu et Travail de civilisation. Médianes, Rouen

Kapferer JN (2012) The luxury strategy: break the rules of marketing to build luxury brands. Kogan Page, London

Kleanthous A (2011) Simply the best is no longer simple—*Raconteur on* sustainable luxury, July 2011, 3. Source: http://theraconteur.co.uk/category/sustainability/sustainable-luxury/ Accessed Dec 2012

Koefoed O, Skov L (unknown year) Sustainability and fashion. In: Openwear: sustainability, openness and P2P production in the world of fashion. Research report of the EDUfashion project. http://openwear.org/data/files/Openwear%20e-book%20final.pdf. Accessed 9 May 2012

Low T (unknown year) Sustainable luxury: a case of strange bedfellows. University of Bedfordshire, Institute for Tourism Research Bedfordshire, Bedfordshire

Marshall J, Coleman G, Reason P (2011) Leadership for sustainability—an action research approach. Greenleaf Publishing Limited, Sheffield

Mortelmans D (2005) Sign values in processes of distinction: the concept of luxury. Semiotica 157–1/4(2005):497–520

Pinkhasov M, Nair RJ (2014) Real luxury: how luxury brands can create value for the long term. Palgrave Macmillan, New York

Popelka CA, Littrell MA (1991) Influence of tourism on handcraft evolution. Ann Tourism Res 18(1):392–413

Rahman S, Yadlapalli A (2015) Sustainable practices in luxury apparel industry. In: Gardetti MA, Muthu SS (eds) Handbook of sustainable luxury textiles and fashion, vol 1. Springer, Singapore, pp 187–211

Ranfgani S, Guercini S (2015) Beyond appearances: the hidden meanings of sustainable luxury. In: Gardetti MA, Muthu SS (eds) Handbook of sustainable luxury textiles and fashion, vol 2. Springer, Singapore, pp 1–16

Romero G (2015) Camino a tenango. Thule Ediciones, Barcelona

Root RA (2005) Introduction. In: Root RA (ed) The latin america fashion reader. Berg, New York, pp 1–13

Scheibel S (unknown year) Ethical luxury—myth or trend? Essay. The London School of Economics and Political Science, London

Sommers C (2014) Pachacuti, UK. In: Gardetti MA, Girón ME (eds) Sustainable luxury and social entrepreneurship: stories from the pioneers. Greenleaf Publishing, Sheffield, pp 71–87

Swain MB (1993) Woman producers of ethnic arts. Ann Tourism Res 20(32):51

Villiger A, Wüstenhagen R, Meyer A (2000) Jenseit der Öko-Nische. Birkhäuser, Basel

Walker S (2006) Sustainable by design—explorations in theory and practice. Earthscan, London

World Commission on Environment and Development—WCED (1987) Our common future. Oxford University Press, Oxford

Fashion(s) from the Northwest Coast: Nuu-chah-nulth Design Iterations

Denise Nicole Green

Abstract Carved and painted onto wood, stone, bone, animal skins, or metal, or woven and knitted into cloth, the material culture from Northwest Coast Native peoples has historically been a one-of-a-kind iteration and a declaration of familial rights and privileges. These items have adorned public and private spaces, including the body, and were traditionally produced by hand. In recent years, some designs have been serialized and mass produced through new technologies such as silk screen and digital printing, adorning everything from coffee mugs to t-shirts, sunglasses, jewelry, and other garments (Roth 2012; Roth 2015). This chapter explores the history of Nuu-chah-nulth First Nations specifically and analyzes their distinctive aesthetics and design practice through the lens of fashion theory. The chapter concludes with a discussion of contemporary Nuu-chah-nulth designers and the circulation of their work. I ask: how does fashion operate within Nuu-chah-nulth social organization and how has ongoing colonialism and hybridization of prestige and capitalist economies transformed Nuu-chah-nulth fashion systems and design ideas? The findings discussed in this chapter draw from ongoing ethnographic research (beginning in October 2009) and archival- and museum-based research at both major and minor institutional repositories in the United States, Canada, Germany, and England.

D.N. Green (✉)
Fashion Design and American Indian Studies, Cornell Costume and Textile Collection,
Cornell University, Ithaca, USA
e-mail: dng22@cornell.edu

© Springer Science+Business Media Singapore 2016
M.A. Gardetti and S.S. Muthu (eds.), *Ethnic Fashion*, Environmental Footprints
and Eco-design of Products and Processes, DOI 10.1007/978-981-10-0765-1_2

1 Introduction

Aesthetics of Northwest Coast Native arts are distinctive: anthropologist Boas (1955) emphasized the role of stylized iconography within the complex social organization of the Northwest Coast. Later, Holm (1965) interpreted the formal elements of design, with emphasis on formline as a critical element in characterization of design from this large region of Aboriginal peoples. Development of design complexity within this region is often attributed to ecological factors, as the Pacific Northwest Coast is rich in sea, land, and sky resources. Indigenous peoples had the luxury of time and abundant natural resources, which enabled them to focus on artistic practice because agriculture and animal husbandry were not necessary on a large scale (Arima and Dewhirst 1990; Duff 1977). As a result, Northwest Coast ceremonial and material culture developed with great complexity and ornate detailing. The potlatch, a ritual occasion to mark important life events through the display and distribution of familial wealth, is one contemporary situation where Nuu-chah-nulth design and fashion are displayed and deployed. In addition to the ceremonial realm, Nuu-chah-nulth graphic artists, weavers, carvers, engravers, and fashion designers are finding commercial opportunities for their work, especially following the popularization of serialized design (e.g., silk screen prints) in the mid-1970s. New forms of cultural expression, including fashion, have burgeoned since the 1970s, supported by a growing art market, governmental initiatives, training programs, increased tourism, and revitalization of cultural practices, such as the potlatch, which had been banned in Canada from 1884 to 1952 (Glass 2008, p. 15).

During the pre-European contact period, before 1774,[1] Nuu-chah-nulth peoples engaged in trade and potlatch celebrations with other Native communities, which fueled fashion change long before European contact. Northwest Coast peoples were very industrious and willing to develop their design practice with new technological knowledge and trade materials that came with trade (Jensen and Sargent 1987; Sproat 1868). By the late eighteenth century, clothing styles changed rapidly as new materials became available (Sproat 1868; Swan 1868). In the late nineteenth and early twentieth centuries, the Canadian government regulated Native dress in numerous ways—from banning the potlatch in 1885 to outlawing Native dress entirely in 1914. It was not until the mid-twentieth century that potlatching was no longer illegal and the last residential schools on the Northwest Coast closed their doors as late as the early 1980s. Over the last 40 years, Northwest Coast Native artists have experienced a cultural revival of sorts. I argue that fashion has always been an integral part of Nuu-chah-nulth ceremonial life, and focus on specific designers who are part of cultural and commercial design efforts.

---

[1]Captain James Cook arrived and landed in 1778, but Juan Pérez anchored off shore in 1774.

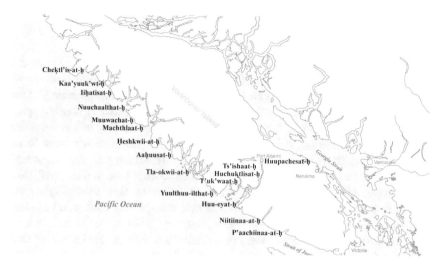

Fig. 1 Map of Nuu-chah-nulth tribal territories along the West Coast of Vancouver Island, British Columbia, Canada. Map created by Chuuchkamalthnii; used with permission from the Belkin Art Gallery

Fashion has existed for the Nuu-chah-nulth as a social and aesthetic form since *Iikhmuut* (time immemorial). The Nuu-chah-nulth hail from the West Coast of Vancouver Island, and are a southern Northwest Coast group that speak a Wakashan language (Fig. 1). The author has spent the last 6 years conducting ethnographic research among Nuu-chah-nulth communities, specifically the Hupacasath and Tseshaht First Nations, but has interviewed highly ranked chiefs and artisans from a number of other nations, including the Ahousaht, Ditidaht, Heshquiaht, Huu-ay-aht, Kyuquot, Cheklesath, and Ucluelet. The author has also conducted research in museums and archives across the United States, Canada, and Europe.[2] This chapter draws upon qualitative ethnographic research, in combination with archival methods, to explore aesthetics, semiotics, material production practices, and circulation of Nuu-chah-nulth fashion design within their prestige economy as well as the emergent capitalist context. I begin by contextualizing the Nuu-chah-nulth historically, then move to interpret Nuu-chah-nulth dress practices through relevant theoretical approaches to fashion, and conclude with a discussion of contemporary Nuu-chah-nulth designers.

[2]Museum and archival collections include: American Philosophical Society, Smithsonian National Museum of Natural History, Smithsonian National Museum of the American Indian, National Anthropological Archives, United States National Archives, Museum of Anthropology at the University of British Columbia, Alberni Valley Museum, British Columbia Provincial Archives, Canadian Museum of History, Royal British Columbia Museum, Cornell Costume and Textile Collection, Karl May Museum, Ethnological Museum of Berlin, British Museum, Fenimore Art Museum, Peabody Harvard, Peabody Essex, Menil Collection, Denver Museum of Nature and Science, and the American Museum of Natural History.

2 Contextualizing Nuu-chah-nulth First Nations

In one of my first meetings with Nasquu-isaqs, a high-ranking and well-respected woman of the Hupacasath First Nation, one of the Nuu-chah-nulth Nations, she explained: "Our people have been here forever, that's our understanding." Days later, standing in front of Kakaw'in, a petroglyph site on *Aa-ukw Tl'ikuulth* (Sproat Lake), Nasquu-isaqs' Uncle, Chuuchkamalthnii, elaborated: "We talk about a time called *Iikhmuut. Iikhmuut* is 'The Time Before Time'. That's clearly prehistoric." Anthropologist, Drucker (1951), among others, has written that the Hupacasath have "made their home on the shores of Sproat Lake since time immemorial" (5).

Anthropologists have repeatedly made mention of this sensibility among Nuu-chah-nulth peoples—that is, deep temporal and spiritual connection to place. In a study of the Kyuquot in the late 1970s, Kenyon (1980) wrote "There is a sense of identification with the land, rooted in the belief that their people have lived there forever and it is a part of their being" (154). Identification with *haahuulthii* (chiefly territories) has both historical and aesthetic depth, in part because of the nature of the landscape. Drucker argued that because the West Coast of Vancouver Island was rugged it "played an important part in the sociopolitical divisions of people" (7). One way to signal sociopolitical divisions is through dress and adornment and early accounts document unique crest symbols, craftsmanship, and the importance of local trade routes (Beaglehole 1999; Boas 1890; Sapir n.d.; Sproat 1868).

Before European contact, Nuu-chah-nulth people lived as autonomous tribes in their particular winter and summer village sites. People migrated with the seasons, women moved to different nations through marriage,[3] warfare affected ownership and occupation of territories, and a complex trade network connected Nuu-chah-nulth tribes as well as other Indigenous populations on Vancouver Island and along the Coast of what is now mainland British Columbia, Alaska, Washington, and perhaps even as far south as Oregon and California. Captain Cook, for example, collected objects that were clearly Aleutian and Tlingit from Nootka Sound, the Nuu-chah-nulth community, where he landed in March of 1778. Although different Northwest Coast tribes have unique approaches to design, Holm (1965) has pointed out many unifying features, such as an "even distribution of weight and movement" through use of formline, ovoids, eyelids, eyebrows, U-forms, horizontal symmetry, semi-angularity of curves, and eyebrows (75). The absence of other features, such as overlapping, results in a two-dimensional and rather flat form, but one that utilizes the entirety of the graphic space with a carefully weighted balance. Although groups share stylistic features, they have distinctive designs which connect to particular histories and rights. Colonial encounters, and later relationships with settlers and industrial economies, would also increase the speed of fashion change in the nineteenth and twentieth centuries.

---

[3]If the woman was of higher social rank than her future husband, the husband would move to the woman's community. *Hiskmiilth* is the term used for a man who moves to his wife's nation.

Encounters between Nuu-chah-nulth and *Maatmalthnii* (White people) on the Northwest Coast began as economic interaction. Trade vessels visited the Coast and later established posts. Russians met with Northern groups as early as 1741, meaning that trade goods made their way to the West Coast of Vancouver Island before European contact (and evidenced by the aforementioned Cook collections at the British Museum). Spaniard Juan Pérez anchored offshore of Nootka Sound in 1774, and Captain James Cook landed near Friendly Cove in March of 1778 and spent a month with Nuu-chah-nulth peoples. By the 1790s, merchant ships from Europe and North America participated in a lucrative sea otter pelt trade. Following the widely-read publication of Cook's journals, a flurry of vessels made the "Golden Round" in the late eighteenth and early nineteenth centuries: they traded for sea otter pelts on the West Coast, brought furs to Canton, China to trade for tea, and eventually returned home to three times the profit. During this period, many ships anchored, encountered, and made transactions with Nuu-chah-nulth people (Gibson 1992).

Cook's journals, among other written accounts, fueled the European imagination of a rugged, wild, untouched landscape. Menzies and Butler (2008) have criticized the notion that the Northwest Coast was an unadulterated landscape that was "tamed" by capitalism and colonialism. These kinds of discourses—that the Coast was "untouched"—became rationale for dispossession. In other words, perpetuating the idea that Indigenous people were not using the land became a justification for taking the land and exploiting resources through extraction industries, agriculture, and eventually settlement. However, in reality, Indigenous peoples throughout the Northwest Coast were active in manipulating terrestrial and marine territories prior to European contact. Nuu-chah-nulth people actively harvested the inner bark of the cedar tree in late spring to be used in weaving or treated and processed to make soft fabrics. Engagement with the landscape and its natural resources enabled and perpetuated the Nuu-chah-nulth fashion system before contact. The earliest drawings of Nuu-chah-nulth people, made by John Webber, the artist aboard Captain Cook's ship, confirm that people were wearing ornate cedar bark clothing, hats with designs made of sea grasses tightly twined (Fig. 2). Examples of elaborately woven and painted cedar bark cloaks, and woven hats with pictorial designs, were collected by Cook and are housed in the British Museum collections.

At the same time that Cook's stories of encounter and the potentiality of wealth made their way back to North America and Europe, sea otter populations began to dwindle and they ultimately disappeared from the West Coast by the mid-1820s. Between the 1830s and the 1860s, very few *Maatmalthnii* came to Vancouver Island for trade. Despite the dearth of outsiders, the *Maatmalthnii* left another kind of violent imprint on place: disease. Although many epidemics went unrecorded, conservative estimates project that by 1881, the Nuu-chah-nulth population declined by two-thirds (Boyd 1990, p. 145). "They wanted us to vanish," Chuuchkamalthnii, a highly regarded contemporary Nuu-chah-nulth designer, often told me during our ethnographic interviews.

Fig. 2 Woman of Nootka Sound, 1778. engraving by W. Sharp from drawing by John Webber, artist aboard the third expedition of Captain James Cook

Nuu-chah-nulth *haahuulthii* and social organization changed dramatically during this period, when sickness and disease dramatically reduced populations on the West Coast. Confederation efforts were emerging just prior to contact (Knight 1996, p. 26), but severely diminished populations sped up the process. The late

Fig. 3 Shawl covered in dentalia shells, likely used for a girl's coming-of-age potlatch. Collected by Captain Dorr F. Tozier in 1902. Smithsonian National Museum of the American Indian, #069293. Photo by Denise Green

nineteenth century brought major transformations to the West Coast of Vancouver Island that would have a tremendous impact on dress practices and access to new materials (Green 2013a): prospecting resulted in mining, logging, and fishing operations that relied on Nuu-chah-nulth labor. Political and governmental changes were also underway: Vancouver Island merged with British Columbia in 1866 to become the Colony of British Columbia. Shortly thereafter British Columbia joined the confederation in 1871, making Native peoples wards of the federal government. Visitors to the West Coast no longer left on their ships, but stayed and began to make their lives there—sometimes intertwined with the Native peoples they met and often came to depend upon.

Lutz (2008) argues that the "moditional economy," which was a hybridization of traditional prestige and emerging capitalist economies, expanded global capitalism through use of Aboriginal labor (279). The moditional economy began with early trade relationships, which fueled an accumulative capitalist system and transformed the landscape, particularly on the West Coast of Vancouver Island. In addition to the landscape, Nuu-chah-nulth fashion also changed dramatically with newfound access to trade goods. As discussed, Nuu-chah-nulth people were initially involved in trade routes because of their access to sea otter pelts. These furs were also highly prized among the Nuu-chah-nulth, and generally only those of chiefly status had the right to don sea otter fur. Display of wealth and status through dress practices persisted after the sea otter's extinction, integrating both local and far-flung materials. Girl's coming of age adornments are a good example: dentalia shells, brass buttons, thimbles, and glass "blue beads" might all be part of the jewelry and embellishments worn for her *Aaytstuulthaa* (Girl's Coming of Age Potlatch). As new wealth came to Nuu-chah-nulth peoples in the form of wage labor, potlatching and visible displays of wealth on the body through fashion became popular (Fig. 3).

Nuu-chah-nulth people shaped capitalist expansion in various ways and they played a critical role in emerging industrial economies (Knight 1996). Lutz (2008) has suggested that Northwest Coast peoples "worked according to a logic that was based in their own economies," with the understanding that they still had access to their own subsistence economies and could work seasonally (203). "Fishing, logging, whaling, hop picking, berry picking, and sealing generated income in the late spring, summer, and early fall months, and allowed Nuu-chah-nulth to accumulate goods in preparation for winter redistribution through the potlatch system... What resulted was a period of wealth and a more upwardly mobile commoner class. Although the elite controlled trade networks, access to wage-labor was a relatively democratic process, resulting in a burgeoning *nouveau riche*" (Green 2013a, p. 171).

The potlatch system, in which families host elaborate gatherings to mark important life events, is a critical component of a prestige economy. The "business"[4] of the potlatch is marked by redistribution of wealth. As a functioning part of a culture that depends upon oral history, potlatch attendees are compensated for their time and commitment of memory and remembering by accepting gifts from the host family. The more the host family gives away to mark their event, the greater their prestige and social status enhances. During the late nineteenth century, wage labor fueled an incredible number of potlatches and enabled many families to become socially mobile. Dress played an important role in shifting social status for Nuu-chah-nulth people of this time.

However, Indigenous economic relations were also undermined through different legal and governmental devices to ensure a labor force for this emerging economy (Menzies and Butler 2008, p. 35). This follows Wolf's (1982) argument that the capitalist market "creates a fiction" of symmetrical relationships—in fact, there was ongoing struggle and negotiation between groups (354). In the latter half of the nineteenth century the government of Canada deployed a series of legal actions on Native peoples that resulted in dispossession of land and encouragement of the industrial economy, which was rapidly transforming the landscape through resource exploitation (e.g., timber, mining, fishing, etc.). Duff (1977) has argued that, unlike other provinces in Canada, British Columbia "evolved a policy that ignored or denied the existence of any native title," and perpetuated this denial through the 1876 Indian Act and other legal actions (67). By the 1880s, Aboriginal people accounted for roughly three-quarters of the population of the province, and it was at this same time that the Aboriginal people were legally denied the right to participate in elections (Lutz 2008). Their voices were silenced in the political realm, and later in the legal realm with the "Great Settlement of 1927," which prevented First Nations people in British Columbia from seeking legal counsel for land claims.

In 1876, the Joint Indian Reserve Commission was set to the task of answering the "land question." The answer was establishment of reserves under the ideological auspice that White settlement was good for Native peoples and would eventually help Natives to "learn how to use land properly" (Harris 2002, p. 108). Peter O'Reilly arrived in 1882 to allocate reserves on the West Coast of Vancouver Island. According to Harris (2002), Premier George Walkem was encouraged, "not to allocate reserves in Barkley Sound if whites had applied for the land, or

---

[4]"Business" is the English term used by Nuu-chah-nulth people to refer to the planned proceedings of the potlatch. If it is a Thlaakt'uulthaa (End-of-grief potlatch), for example, the "business" would be to end and put away the family's grief for the death of a family member publicly. Giving names is also another form of "business" that often takes place at potlatches.

simply if Indians used a site for fishing. The premier was afraid that the best harbors would be lost" (207). Such large-scale legal, institutional, governmental, and economic encounters were part of "power geometries of space" which produced new relationships to landscapes and decreased access to natural resources used for the production of textiles, dress, and jewelry (Massey 1994, 2005).

Race-based education also fueled inequalities and impeded intergenerational knowledge transfer around weaving, design, and semiotics of dress. Indian Residential Schools (IRS) provided sub-par education that attempted to assimilate that of the Nuu-chah-nulth children. In my interviews with residential school survivors, many were quick to point out that uniforms and short haircuts were required. IRS regulated bodies and minds, and sought to modify the appearances of Native children and to prevent students from learning about their cultural practices. In 1920 an amendment to the Canadian Indian Act of 1876 made residential school attendance mandatory for Native children. Nuu-chah-nulth children were alienated from their homes, forced to speak English, and were not allowed to practice cultural traditions or wear clothing, jewelry, or hairstyles that appeared "Aboriginal," as were other First Nations youth across the country. In Nuu-chah-nulth territories, The Alberni Indian Residential School operated until 1973 and the Christie (Kakawis) Indian Residential School did not close until 1983, making it the last residential school in operation in British Columbia.

Early encounters between Nuu-chah-nulth and visitors stimulated trade, but as visitors became settlers, *Maatmalthnii* sought to change Aboriginal cultural lives (Duff 1977). As a result, Duff (1977) has argued "Indian cultures ceased to function as effective integrated systems of living" (53). Anthropologists felt a need to document what they called "vanishing cultures." Edward Sapir worked on behalf of the government of Canada to document the Nuu-chah-nulth of the Alberni Valley in 1910 and 1913–1914 (Sapir n.d.; Sapir and Swadesh 1939, 1955). His student, Morris Swadesh, continued his work in the late 1930s, and Susan Golla picked it up in 1976–1992 (Golla 1987). The Nuu-chah-nulth have also been well documented in museum collections around the world. From the British Museum (Captain Cook collections) to the Ethnological Museum of Berlin to the Smithsonian collections and the American Museum of Natural History, to name a few, Nuu-chah-nulth material culture has left Canada and has been dispersed across continents.

The Northwest Coast became a landscape where different kinds of relationships to the land and to aesthetic and cultural practices were in conflict. On the one hand, accumulative capitalism sought to extract as much from the land as possible, and employed mostly Aboriginal labor to do so (particularly between 1860 and 1920). These practices conflicted with the way Aboriginal relationships with the land have been described in ethnographic texts. Knight (1996) has argued that industrial resource extraction differed greatly from traditional hunting, gathering and processing methods (9). These beliefs about landscape, coupled with participation in the resource extraction industry, would produce a highly conflicted experience of landscape for Nuu-chah-nulth peoples. As a result, the resources used for dress and textiles began to shift and change as new materials became available and

Nuu-chah-nulth people had increasing access to capital and decreasing access to their territories. The adornment of the body began to change, as did the purposes of adornment.

Most recently, the colonial experience has come to a head in British Columbia with the modern treaty process. In 1997, *Delgamuukw v. British Columbia* ruled that Aboriginal title in most of British Columbia had not been extinguished (with the exception of the Douglas Treaties on Vancouver Island). The modern treaty process is a political means of "solving" the "land question." Treaties, called "final agreements," once ratified and implemented, extinguish most right and title to lands. In those treaties that have been implemented, on average, First Nations maintain title over 5–10 % of their traditional territories. Ultimately, the "land question" and relationships to particular places are produced and reified through bodily adornment; therefore, historical contextualization of the West Coast of Vancouver Island is critical to understanding how the fashion of Nuu-chah-nulth people has not only changed but also reflected political, economic, and social adaptations that have come with the colonial encounter.

Fashion is a product of articulating economic, political, cultural, and aesthetic worlds and change with time and effects of increasing globalization. The contemporary ceremonial and commercial fashion practices on the West Coast of Vancouver Island have been produced by nearly 250 years of colonial encounter, exchange, tension, and resistance. I have prefaced the theoretical discussion with an explanation of Nuu-chah-nulth social organization alongside historical shifts in economic systems, colonial regulations, and White settlement on the Pacific Northwest Coast. I now move to discuss theoretical approaches to fashion studies, and how Nuu-chah-nulth articulations of dress as fashion intersect with the aforementioned histories.

3 Theorizing Indigenous Fashion: Phenomenology, Place, and Change

Fashion—that is, the changing stylization of the body over time—is a process that occurs in a variety of manifestations across all cultures. Kaiser et al. (1995) define fashion as "a dynamic phenomenon that inextricably links aesthetics, culture, economics, and everyday social life" (172). Although fashion is affected, in part, by economic systems, it is not exclusively "the product and domain of Western capitalism" (Lillethun et al. 2012, p. 75). Lillethun et al. argue that the Eurocentric definition of fashion, which has come into popular use colloquially and scholastically, was produced by the capitalist and Social Darwinist context of late nineteenth century: "For them fashion occurred in a capitalist production system of innovation, distribution, and consumption wherein the social structure enabled, even fostered, emulation of adjacent status groups" (75). I follow Lillethun et al.'s contention that fashion is not a uniquely Western phenomenon but one that transpires in cultures throughout the world, from Aboriginal to European, rural to urban, and capitalist to communist. Fashion operates differently in each situation, depending on distinct cultural identities, historical contexts, politics, economics, and access to resources. For the Nuu-chah-nulth, fashion pre-European contact unfolded through a prestige economy and in conversation with other Native communities connected through trade routes. With contact and European encounters, the prestige economy began to hybridize with wage labor capitalism, particularly in the late nineteenth century, and continues to this day to function in conversation with late Western capitalism and relationships with *haahuulthii* (chiefly territories and their resources).

Fashion is a powerful material, aesthetic, and bodily force that transforms identities, experiences, and inter/intra-cultural interactions. This is why anthropologist Tarlo (1996) has written that the study of clothing and dress is a critically important anthropological pursuit. On the Northwest Coast, and within Nuu-chah-nulth communities specifically, the transformational power of regalia lies in its ability to connect spiritual and terrestrial worlds. Chuuchkamalthnii (then, Ki-ke-in/Ron Hamilton) has written about oral stories describing the transformational power of dance robes (Jensen and Sargent 1987, p. 64). Kwatyaat, a mythical character, could adorn himself in robes and, in so doing, become a baleen whale. In another story, a group of young boys put on robes that transform them into dogs and once the robes are burned the boys quickly mature into expert whalers (Jensen and Sargent 1987, p. 64; Sapir and Swadesh 1939). The belief in the transformational role of clothing in Nuu-chah-nulth culture is crucial—there is no more powerful force than that from the adornment of regalia in ceremonial settings. Even in the secular environment, the display of crests and images on everyday clothing items such as screen-printed t-shirts, sunglasses, socks, and hats, alludes to the power of graphic crests to transform and strengthen the wearer.

There are two critical forces in Nuu-chah-nulth dress that emerged throughout my fieldwork research: (1) the ability of fashion to connect and make declarations about place and territory; (2) the important role of the body as an intimate site of cultural production and transformation. These findings have guided me toward theories of phenomenology, space and place, and cultural studies of fashion to interrogate the body as a site where identities and relationships to place are produced, and, at the same time, struggles and conflicts unfold in the ongoing colonial context.

Bodies come together to produce spaces and identities, and meanings are negotiated through interactions. For example, what might be called "fashions" are the resonance of multiple bodies referencing similar aesthetics, and when brought together in multiples, produce strong political, ideological, and/or identity-based claims. In developing a theoretical framework for interpreting Indigenous fashion, I begin with an overview of phenomenological perspectives on the body and spatial production that leads into a discussion of symbolic interactionist approaches to fashion theory. I discuss how Nuu-chah-nulth fashioned bodies have been mobilized and regulated across different power struggles, and throughout draw on examples of textile production, bodily modification, appearance management, and fashion as social, aesthetic, and symbolic processes that produce spaces, identities, and new commercialized fashion products.

3.1 Phenomenological Interpretations

Phenomenological approaches to the study of space and place begin with perception. According to Merleau-Ponty (1962, p. 240), the body is where sensation occurs and perception begins. He argues that the body is not a thing in space, but rather, it is of space (171–173). In other words, the two are mutually constitutive—space cannot exist without bodies, and bodies cannot exist without space. This is why he suggests that "all senses are spatial" (252). Fashion, necessarily, comes from and produces spaces through an embodied subject. Casey (1996) has explained that the phenomenological turn from the body-as-object to the body-as-lived has opened up possibilities for a "corporeal subject who lives in a place through perception" (22). For Casey, perception is rooted in corporeality and the mind-body makes places through direct interaction in the immediacy of the present moment. If bodies actively make places, as phenomenologists suggest, the adornments and bodily modifications profoundly connect, and help to produce, senses of place. This body of theory is important to the interpretation of Indigenous fashion, which by being rooted is connected to places of origin—Indigenous peoples are First Nations peoples. The sensorial body is one means of articulating and negotiating ongoing and changing relationships to space.

Casey (1996) builds on the work of Merleau-Ponty, contending that place differs from space in that it is localized and particular to embodied experiences. For example, Nuu-chah-nulth *huulthin* (secular and ceremonial shawls) were historically made from beaten and processed cedar bark, harvested from the forests along the West Coast of Vancouver Island. As Nuu-chah-nulth relationships to these places shifted with industrialization and colonial exploitation of timber resources, so too did fashions change (Green 2011, 2013a). Nuu-chah-nulth people today have trouble accessing large stands of timber because clear cutting has destroyed most forests easily accessible by road (Green 2011). Casey (1996) argues that bodies come to reflect the places they occupy—and the absence of cedar bark clothing on Nuu-chah-nulth bodies reflects destruction of forests and a forced shift to capitalist economies. Particular knowledge about cedar processing, which enabled the production of a bark shawl softer than any wool, was destroyed by the residential school system (i.e., forced removal of children from networks of intergenerational knowledge exchange) and impeded by industrial logging operations. Nuu-chah-nulth have struggled, through the courts and through daily lives, with the government's choices about logging tenure rights. Although some nations today operate logging operations, most logging tenure rights are awarded to large corporations. These political and economic decisions ultimately alienate Nuu-chah-nulth peoples from their traditional resources. As a result, dress practices have changed as people witness their natural resources being annihilated through logging. The phenomenological experience of gathering cedar bark, processing it, and wearing it against the body has been undermined by environmental destruction.

3.2 Phenomenology and Symbolic Interaction

Other scholars have integrated phenomenology and theories of space and place to consider the impact of individual bodies as social bodies. Richardson (1982) follows the phenomenological contention that we are not external to the world, but necessarily part of it. He diverges slightly in his analysis of space by following symbolic interaction, which posits that people act toward objects, places, and one another, based upon the meanings that those people, places, and/or things hold for them (422). His study of two distinct places in Costa Rica—the market and the plaza—found that people altered their behaviors and bodily dispositions upon entering the new space and engaging with the materials found there. This is not unlike the findings of Goffman (1959) over 20 years earlier, who argued that in Western society people are always moving between "front" and "back" stages— that is, between formal and informal spaces of interaction. People alter their behavior depending on which part of the stage they find themselves. He calls front stage performances dramaturgical, where people consciously embody dispositions, mannerisms, appearances, and behaviors that follow perceived social values and ideals and conceal those behaviors that are not in line with an idealized conception of self. Goffman argues that moments of transition are hugely informative because they reveal the putting on and taking off of airs. How people perceive social spaces affects how their bodies perform in those spaces. The potlatch ceremony is one of these spaces, where a literal front and back stage exists, delineated by a *thli-itsapilthim* (painted ceremonial cloth screen). Ceremonial regalia is adorned, and family members and close friends dance onto the floor together, unified as a family or dance group. The adornments enable transformation: the headdress, shawl, and sometimes mask or other implement held in the hands (i.e., feather, paddle, spear, etc. depending on the dance) change the wearer from an individual to a family member. Those in attendance as witnesses watch carefully and quietly, giving full attention to the *ha'wilthmis* (chiefly treasures) coming alive through dance and adornment.

The foundation of Symbolic Interaction theory posits that, in social interactions, people look to one another to try to understand the others' actions. In Nuu-chah-nulth potlatch dances, the family communicates to the witnesses present through song, dance, speech making, and adornments. Dance shawls and ceremonial cloth screens become a medium for crest display. They declare what rights and privileges each family owns, making fashion objects an important form of familial patrimony. The potlatch, with all its pomp and circumstance, is a space for display and communication. Other families adorn their regalia and take the potlatch floor in song and dance to support the host family and whatever business they are conducting. Potlatches continue today as spaces for families to honor their ancestors and display connections via embodiment of dress through dance.

Although late modern capitalism yields fast fashion in the secular world, it yields a slow and dynamic fashion system in the ceremonial world. This does not mean that ceremonial fashion stagnates: on the contrary, people invest immense

amounts of time designing and constructing regalia and it changes with time
and experience. The abundance of new materials, made possible by the capitalist system, also fosters new design. Each Nuu-chah-nulth family relies on some
talented relative, and there are often many. Designers work in a variety of media—
wood carving, graphic design, sewing, metal, and weaving—and are not afraid to
embrace new technologies such as screen printing and new materials such as ultra
suede, acrylic paints, polyester fabrics, synthetic dyes, etc.

Dance shawls typically worn by women are called *huulthin*. They are probably
the most visible and declarative form of regalia. There are two classes of *huulthin*:
individual and familial. Women typically own a personal shawl which they do not
share with anyone else. This is a distinctive feature of Nuu-chah-nulth dress, as
relative groups such as the Kwakwakw'akw have shawls that are shared. An individual shawl is typically constructed by the owner with crests that they or their
family own. Some women may also own individual shawls that have been gifted
to them. The second class of shawls, familial, are typically owned collectively by
the family to be used for specific dances. For example, the sisters of Sayach'apis,
Tayii (head chief) of the Nash'asat-h Tribe, made shawls with the *naani* (grizzly
bear) crest after their brother's near death experience with a grizzly bear. During
his *Yaxmalthit* (cleansing potlatch), his family ceremonially brushed *kahmis*
(deathness) off of his body and then danced with the new crest, which was created
by artist Ray Sim (Fig. 4). The men of the family wore vests screen printed with

Fig. 4 *Naanii* (grizzly bear) crest design for Sayach'apis (Walter Thomas), *Tayii* (chief) of the
Tseshaht First Nation. Shawls were displayed for his *Yaxmalthit* (Cleansing Potlatch), held in
February of 2010. Photo by Denise Green

the same design. The women wore appliqued shawls featuring the same design. As potlatch hosts, the family appeared unified and carried the message of their brother's survival through the grizzly crest. At the end of the potlatch, Sayach'apis was seated in his father's position, making him the head of an important Tseshaht family and tribe.

From unconscious, and often repetitive expressions of the body, particular identities come into being. Bourdieu (1977) calls this outcome *habitus*, the combination of a performative body and cultivated "taste" or aesthetics. Potlatches are a mechanism for intergenerational transfer of knowledge. Typically, a family meets weekly before a potlatch to prepare for dances and songs. Women spend hours in sewing rooms constructing regalia. All of these preparatory meetings away from the potlatch are also opportunities for knowledge transfer within a family. The potlatch itself is an opportunity to share with a larger network of families with their own distinct cultural patrimony and practices. In his book *Distinction*, Bourdieu (1984) explores how a "class habitus" emerges through the body. He argues that "embodied cultural capital of the previous generation" is imparted on youth, generally within familial situations (70, 71). Children develop classificatory realms and the ability to make judgments between these realms—that is, the deployment of "taste" (170). Nuu-chah-nulth social organization is incredibly complex and hierarchical. Potlatches provide an opportunity to display wealth and become socially mobile. A family's prestige is elevated by giving away wealth at the end of the potlatch, and by displaying wealth (in the form of fashion, dances, songs, etc.). The preparations prior to a potlatch also enable children and other family members to learn and practice bodily movement (i.e., dance) and adornment (i.e., fashion).

3.3 Fashion, Colonialism and Regulation of the Body

Fashion change also occurs through regulation of the body and dress. Regulation of appearance and discipline of bodies has been crucial to the colonial project (Comaroff and Comaroff 1997). On the Northwest Coast of Canada, First Nations have been subjected to a body of law called the Canadian Indian Act. Butler (2006) builds upon the work of Foucault (1995) to argue that bodies are sites of discipline and regulation (183–185). Some bodies are more regulated than others, and this is especially evident with aggressive assimilationist policies in the United States and Canada. Because the body surface is relatively variable (i.e., changeable), it easily becomes a contested site of political regulation. Take, for example, the 1914 amendment to the Canadian Indian Act, which required First Nations to gain permission from government officials before appearing in public wearing any kind of ceremonial regalia. This amendment created a body of law that regulated how Aboriginal bodies in Canada could and could not look, and was a means of perpetuating an agenda of assimilation. These actions affected public appearances, and created an even greater difference between secular and ceremonial fashion.

3.4 Dialectical Fashion and Symbolic Interaction

The body is a highly contested and sought-after site—politically, morally, ideolog-
ically, economically—and therefore it is no wonder that understanding/performing
one's identity through fashion is wrought with tension and anxiety (Kaiser 2001;
Davis 1992). Tension may be destructive or productive, and Fred Davis argues
that tensions between competing or conflicting identities are central to promulgat-
ing fashion change. Tensions have many forms for Indigenous peoples from the
West Coast of Vancouver Island: tensions between past and present, traditional
and new materials, ceremonial and secular aesthetics, and commercial markets and
ceremonial outlets for fashion items. Cultural tensions are not post-contact phenomena. Certainly before contact, as evidenced in oral recounting at contemporary
potlatches, families have historically (as well as today) disagreed about ownership, imagery, and histories. Potlatches themselves are essentially about tension
between life phases, often marking an important moment of tension and resolution, such as a birth, coming-of-age, marriage, and death (end-of-grief).

The dialectical approach to fashion holds that change arises from tensions within
a particular culture. People often seek resolution of tension through appearance and
adornment. Turner (1980) found that with the Kayapo of the Amazon, the feeling
of "being in fashion" gave greater security than religion (114). Symbolic interactionist Blumer (1969) has argued that the sense of "being in fashion" and of collectively selecting particular styles is a coping mechanism that allows people to
adjust collectively to a rapidly changing world. Museum artifacts collected in the
late nineteenth century show that Nuu-chah-nulth people were increasingly integrating new materials into ceremonial paraphernalia. From headdress paints derived
from Reckitt's Laundry Blueing to shawls woven with strips of trade blankets,
new materials were integrated into new designs. "The old costume of the natives
was the same as at present, but the material was different" wrote Indian Reserve
Commissioner Gilbert Malcolm Sproat in 1868. Evidence in museum collections
suggests a rapid and consistent adoption of new materials in the late nineteenth century, integrated with historically used materials such as cedar bark and sea grasses.

The dialectical approach to fashion change also relies on the fact that "fashion susceptible instabilities," particularly those related to identity, continuously
find resolution and tension once more. People's adoption of similar fashions
at one moment in time displays unity and synthesis, but latent and lurking tensions arise with everyday matters and world events, and the dialectic continues
to unfold. Changes in appearance follow identity tensions and resolution (Davis
1992). The argument is that identity tensions and conflicts may be "read" through
fashion changes over time. One such tension is that between youth and adulthood.
With menarche, Nuu-chah-nulth women are isolated and later revealed to the family through an *Aaytstuulthaa* (Girl's Puberty Potlatch). Her dress has completely
changed, she wears ear pendants and some representation of wealth (traditionally
dentalia shells, which may still be used, but more typically, dollar bills are pinned
to her dance shawl). The elaborate costuming, different from anything she has

worn publicly, signals the result of identity tension between youth and adulthood. The synthesis is an adult dress. She likely abandons her childhood dance shawl, takes a new name, and displays her new identity through dress.

One of the greatest tensions has been around the commercialization of Nuu-chah-nulth designs. As discussed, crests are familial patrimony and clearly owned, and have thus far only been described and interpreted within the ceremonial life of Nuu-chah-nulth people. The reality for Nuu-chah-nulth designers today is that they must grapple with family, tribal, and national politics, whilst the larger fashion industry continually appropriates from Native culture and design. The emergence of new technologies, such as silk screen printing, has enabled serialization of images and increased distribution. I conclude by discussing eight contemporary Nuu-chah-nulth designers who work in a variety of media which ultimately are used as adornment on bodies.

4 Nuu-chah-nulth Fashions in the Contemporary Context

The translation of Nuu-chah-nulth design into the commercial realm really began with curiosities and souvenirs for early trade vessels and late nineteenth century tourists and the liquidation of ceremonial paraphernalia by art dealers, anthropologists and government officials from the late nineteenth century (Cole 1985). Northwest Coast art and design was appreciated by traders, tourists, and later settlers, and, despite aggressive assimilation efforts by the Canadian government, Northwest Coast Native artists continued to work throughout the potlatch ban and residential school epoch. The contemporary Northwest Coast Art market began to thrive in the 1970s with the Northwest Coast Indian Artists Guild, apprenticeship programs such as Thunderbird Park at the Royal British Columbia Museum, and the continued devotion of anthropologists and art historians to the documentation and celebration of Northwest Coast arts in public institutions. The proliferation and celebration of Northwest Coast art has created an interesting space for commercialization of decorative arts and fashion. Jewelry, both metal and basketry, was an early form of fashion sold to outsiders. With basketry and weaving talents, hats were also a popular tourist item, and Fig. 5 is a great example of combining Nuu-chah-nulth wrapped twining techniques, local materials, with the fashion desires of a Western consumer in the late nineteenth century. Lastly, silk screen prints have become an important part of contemporary fashion design. There are many hundreds of Nuu-chah-nulth designers that work in both large and small scale.

4.1 Lena Jumbo (Ahousaht)

Lena Jumbo (Fig. 6) is one of the leading weavers on the West Coast of Vancouver Island and lives at Maaqtusiis, the Ahousaht village a short boat ride from Tofino. Lena was born in the 1930s and learned to weave from her grandmother at a

Fig. 5 Basketry grass boater hat produced in the late nineteenth century and collected by James G. Swan for the Smithsonian National Museum of Natural History. (E23330) Department of Anthropology, Smithsonian Institution. Photo by Denise Green

Fig. 6 Photograph of Lena Jumbo holding a whaler's hat in 2010. Photograph by Denise Green

young age (her mother died when she was 3 years old and her grandmother looked after her). Today, Lena produces a range of work: small *pika-uu* (curio style baskets), traditional whaler's hats, shopping baskets, pendants, and various implements covered in basketry (e.g., bottles, cigarette lighters, glass buoys, etc.). Her best sellers, however, are miniature whaler's hats worn as earrings (Fig. 7). With the exception of synthetic dyes and commercially manufactured sterling silver fish hook earrings, Lena does not employ any other new technologies in her weaving. Everything is still produced by hand using materials gathered locally such as cedar bark and sea grasses. The designs "are all in my head," Lena has told me, "I don't use a pattern book."

Fig. 7 Miniature whaler's hat earrings made by Lena Jumbo in 2010 featuring a whale design. Photograph by Denise Green

4.2 Chuuchkamalthnii (Ki-ke-in/Ron Hamilton—Hupacasath)

Chuuchkamalthnii is one of the most prolific contemporary Nuu-chah-nulth designers, though today he designs exclusively for ceremonial events only (Fig. 8). He is adamant that his work is not "art." "There's a lot of literature out there that talks about the 'Indian artist, Ron Hamilton.' Well, I'm not an Indian and I'm not an artist. Never have been" (Green and Chuuchkamalthnii 2010). Chuuchkamalthnii resists labels of all kinds, the subtext being that these are English words and Western categories that do not do justice to the spiritual and cultural significance of his design work. Chuuchkamalthnii was born in 1948, and as a young person he engaged with the commercial market. He sold drawings, carvings, silver engravings, and silk screen prints (Fig. 9). From 1969 to 1973 he worked at Thunderbird Park at the Provincial Museum of British Columbia as a carver, restoring and replicating old totem poles and creating new original works. During this time he was apprenticed with the renowned Kwakwaka'wakw carver Henry Hunt. He later co-founded the Northwest Coast Indian Artists Guild in 1972, with the intention of supporting and teaching young Native designers interested in creating new work for both commercial and ceremonial settings. The British Museum, the Museum of Anthropology at UBC, Port Alberni's Rollin Arts Centre, Public Archives of British Columbia, the Belkin Art Gallery, and the Alberni Valley Museum have all featured retrospectives of his printmaking,

Fig. 8 Photograph of Chuuchkamalthnii painting a dance shawl in 2012. Photograph by Denise Green

Fig. 9 Chuuchkamalthnii preparing a silk screen print in the late 1970s. Photo used with permission of Chuuchkamalthnii

poetry, photography, and drawing. Examples of his carving and other design work are found in the Canadian Museum of History, the Museum of Anthropology at the University of British Columbia, the Alberni Valley Museum, the Royal British Columbia Museum, and the British Museum, among others. Chuuchkamalthnii's resistance to the commercial market and capitalist exploitation of Northwest Coast design has left him devoted to the fashions of potlatching and ceremonial life. He regularly produces designs for dance shawls, ceremonial curtains, potlatch t-shirts, and recently completed carving a totem pole for the Ts'uubaa-asatx (Lake Cowichan First Nation). He occasionally carves bone pendants and engraves metal, but more typically focuses his efforts on shawls and graphic design. Chuuchkamalthnii is devoted to the meaning of these images, and is careful to follow his ancestors' teachings and stories in creating appropriate designs for particular families. He uses his immense knowledge of history and Nuu-chah-nulth social organization to ensure that the design work he completes for ceremonial purposes has been vetted and thoughtfully considered before it appears on the potlatch floor.

4.3 Joe David (Tla-o-qui-aht/Hupacasath)

Joe David's father was Tla-o-qui-aht and his mother was Hupacasath, but he was born in the Tla-o-qui-aht village of Opitsaht. He spent much of his youth and adult life in Seattle. David is a first cousin to Chuuchkamalthnii, and born just 2 years before him. Both were members of the Northwest Coast Indian Artists Guild, but whereas Chuuchkamalthnii has chosen to focus on ceremonial life and inter-community fashion, David has succeeded in the commercial market. His designs have been sold as one-of-a-kind art pieces, but more commonly as serigraphs. The serigraphs of the 1970s influenced the development of the Northwest Coast art-ware market, which has continued to grow since the 1980s (Roth 2015). Although David's work is more likely to be found in art galleries, his serigraph designs paved the way for Northwest Coast artware markets and the appearance of silk screen prints on garments such as t-shirts, sweaters, and hoodies.

4.4 Tim Paul (Hesquiaht)

Tim Paul, born in 1950, is a designer of the same era as David and Chuuchkamalthnii. As did Chuuchkamalthnii, he participated in the Thunderbird Park carving apprenticeship program and was ranked "First Carver" at the Royal British Columbia Museum until 1992. Although Tim Paul is known for his carving talents, he has also embraced clothing design and produced silk screen prints for garments sold in the commercial market (Fig. 10). The design shows a *hinkiits* dancer wearing an elaborately designed robe and is fascinating because it depicts a ceremonial right that is often jealously guarded.

Fig. 10 Tim Paul's silk
screen print design for
a hoodie, circa 2012.
Photograph by Denise Green

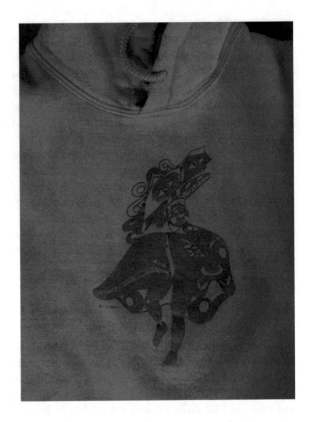

4.5 Julie Joseph (Ditidaht)

Julie Joseph is known for her fine basketry. Similar to Lena Jumbo, she uses her
skilled hands to create pendants and earrings, transforming weaving practices typically used for millinery and baskets into the realm of jewelry design. Joseph grew
up in a traditional Ditidaht village site with her immediate family. The rest of the
community had moved to a new village site, Balaatsad, which had electricity, run-
ning water, and road accessibility. She remembered learning to weave by coal oil
lamp, and has taught her children to weave. In her adult life she has lived in larger
cities such as Nanaimo, where there is a market for her jewelry designs (Fig. 11).
Joseph adheres to the designs owned by her family—that is, whales, canoes, and
harpoons—because her ancestors were whalers (Green 2013b).

4.6 Paul Sam (Ahousaht)

Unlike the other designers discussed thus far, Paul Sam focuses primarily on metal
engraving and produces jewelry for the commercial market (Figs. 12 and 13). As with

Fig. 11 Earrings designed by Julie Joseph. Photograph by Denise Green

Fig. 12 Silver cuff design featuring *hinkiist'am* (serpent headdresses), made in 2010 by Paul Sam. Photograph by Denise Green

Fig. 13 *Hinkiits'am* pin designed by Paul Sam circa 2008. Photograph by Denise Green

Tim Paul's hoodie design, Sam also uses ceremonial imagery such as *hinkiits'am* (serpent headdresses). He carves deep grooves into silver and uses cross-hatching and dashing to create texture and value.

4.7 Yahmiss—Jolleen Dick (Hupacasath)

Born in 1991, Yahmiss is the youngest designer included in this discussion. She translates traditional Nuu-chah-nulth basketry designs into woven beadwork (Fig. 14), almost exclusively drawing from the basketry designs of her great grandmother, Rose Cootes (Ucluelet). It is important to Yahmiss that her designs reflect family patrimony and the rights that come to her through her maternal grandfather, who was the hereditary chief of the Hupacasath. Yahmiss majored in Aboriginal Tourism at Vancouver Island University, and currently serves on the council for the Hupacasath First Nation. During the summer months, she organizes a Wednesday night market at Victoria Quay in Port Alberni, thus providing artisans a space to sell their design work. She is interested in both the design and marketing aspects of fashion, and actively works to bring Aboriginal tourism and Nuu-chah-nulth fashion design together in the commercial market.

Fig. 14 Beaded bracelet, woven using peyote stitch and Thunderbird design by the late Rose Cootes, 2013. Photo by Denise Green

4.8 Tl'aakwaa Huupalth—Carmen Thompson (Ditidaht/Kyuquot)

Of all Nuu-chah-nulth designers, Carmen Thompson is probably the most directly involved with the fashion industry and also works as a costume designer in the film industry. Thompson is the daughter of Art Thompson, a renowned Ditidaht artist known for his carving, painting, and serigraphs, who passed away in 2003. Carmen Thompson was inspired by her father and had talked with him about her ideas to create a Nuu-chah-nulth fashion line. She received a scholarship and attended the Fashion Institute of Design and Merchandising in Los Angeles, and eventually began to work as a freelance costume designer in Hollywood. She is currently working for Aarrow Productions on costumes for *1491: The Untold Story of the Americas Before Columbus*. Since 1997, Thompson has operated a design company called TlaKwa (meaning "copper," suggesting wealth). In 2013 she produced 60 hoodies stenciled with her father's design for the Indian Residential School Survivors Society (Fig. 15). Thompson uses new technologies such as Photoshop and the internet both to create and deploy her design work. Thompson is an up-and-coming Nuu-chah-nulth fashion designer and costumer and can certainly spearhead the future of Nuu-chah-nulth fashion design.

Fig. 15 Hand stenciled hoodie made by Carmen Thompson, using a design for the Indian Residential School Survivors Society produced by her father, Art Thompson. Created for the Indian Residential School Survivors Society 2013 Youth Conference. Photo by Carmen Thompson

5 Conclusion

Nuu-chah-nulth peoples participated in a dynamic and highly developed fashion system prior to European contact. Fashion was never stagnant or fixed, but changed with seasons, interactions with other communities (including trade/exchange), and internal social dynamics. Nuu-chah-nulth fashion change was therefore fueled by the prestige economy, highly developed ceremonial life, and hierarchical social organization. Easy access to an array of abundant natural resources and trade networks with other First Nations facilitated exchange of ideas, aesthetics, and materials. Materials used to produce fashion items were both locally sourced and exchanged: Nuu-chah-nulth people used cedar bark, animal skins, dentalia shells, and sea grasses to produce early garments, and traded for mountain goat hair, dog hair, and other trade items. These natural resources were completely biodegradable and left a very small impact on the environment. After European contact, the Nuu-chah-nulth fashion system changed rapidly with an emerging capitalist economy, wage labor, and governmental regulation of Native bodies. As trade networks became more expansive, new materials and resources fueled dramatic changes in style and made fashions much less sustainable. Governmental regulations required Nuu-chah-nulth people to adopt "Western" dress in 1914 and this meant purchasing fabrics and other materials from outside sources, making Nuu-chah-nulth design much less sustainable both socially and environmentally. As governmental regulations loosened in the late twentieth century, a revitalization of Native art and design emerged alongside new technologies such as serigraphs. Contemporary designers draw upon traditional skills, such as carving and weaving, to create new designs that appear in mass-produced fashion as well as one-of-a-kind originals in art galleries and museums. Although some designers have returned to using natural resources from the region, many continue to source fabrics, paints/printing inks/dyes, and other materials from outside sources.

Fashion theory has tended to favor Western design as "fashion" and non-Western, Indigenous design as "costume," but these categories began to change with post-modern, post-colonial thinking in the 1980s and 1990s. This chapter illustrates how fashion theories derived from phenomenology, Symbolic Interaction, anthropology, and cultural studies, in combination with Indigenous thought and reflection through ethnographic practice, may be used to interpret and understand the complexities of Indigenous fashions past and present.

References

Arima EY, Dewhirst J (1990) Nootkans of Vancouver Island. In: Suttles W, Sturtevant W (ed) Handbook of North American Indians, vol 7, Northwest Coast. Smithsonian Institution, Washington, DC, pp 391–411

Beaglehole JC (ed) (1999) The journals of captain James Cook on his voyages of discovery, edited from the original manuscripts. Boydell Press Hordern House, Woodbridge

Blumer H (1969) Fashion: from class differentiation to collective selection. Sociol Q 10(3):275–291

Boas F (1890) Second general report on the Indians of British Columbia. British Association for the Advancement of Science, 60th meeting, pp 572–715

Boas F (1955) Primitive art. Dover Publishing, New York

Bourdieu P (1977) Outline of a theory of practice (trans: Nice R). Cambridge University Press, Cambridge

Bourdieu P (1984) Distinction: a social critique of the judgement of taste. Harvard University Press, Cambridge

Boyd RT (1990) Demographic history: 1774–1874. In: Suttles W (ed) Handbook of North American Indians, vol 7. Smithsonian Institution, Washington, DC, pp 135–148

Butler J (2006) Gender trouble: feminism and the subversion of identity. Routledge, New York

Casey E (1996) How to get from space to place in a fairly short stretch of time: phenomenological prolegomena. In: Feld S, Basso K (eds) Senses of place. School of American Research Press, Santa Fe, pp 13–52

Clayton DW (2000) Islands of truth: the imperial fashioning of Vancouver Island. University of British Columbia Press, Vancouver

Cole D (1995) [1985] Captured heritage: the scramble for Northwest Coast artifacts. University of British Columbia Press, Vancouver

Comaroff J, Comaroff J (1997) Of revelation and revolution: the dialectics of modernity on a South African frontier. University of Chicago Press, Chicago

David J et al (1977) Northwest Coast Indian artists guild, 1977 graphics collection. Canadian Indian Marketing Services, Ottawa

Davis F (1992) Fashion, culture, and identity. University of Chicago Press, Chicago

Drucker P (1951) The northern and central Nootkan tribes. For sale by the Supt. of Docs., U.S. Govt. Print. Off, Washington

Duff W (1977) The Indian History of British Columbia: the impact of the white man. In: British Columbia Memoir No. 5. British Columbia Provincial Museum, Victoria

Dunae PA (n.d.) viHistory. http://vihistory.ca/index.php. Date visited: 19 Feb 2013

Foucault M (1995) Discipline and punish: the birth of the prison. Vintage Books, New York

Gibson JR (1992) Otter skins, boston ships, and China goods: the maritime fur trade of the Northwest Coast. University of Washington Press, Seattle, pp 1785–1841

Glass A (2008) Crests on cotton: "Souvenir" T-shirts and the materiality of remembrance among the Kwakwaka'wakw of British Columbia. Mus Anthropol 31(1):1–17

Goffman E (1959) The presentation of self in everyday life. Doubleday, Garden City

Golla S (1987) He has a name: history and social structure among the Indians of Western Vancouver Island. Doctoral Dissertation, Columbia University, New York

Green DN (2011) Mamuu—La Pratique du Tissage/Mamuu—The practice of weaving. Cahiers métiers d'art/Craft J 5(1):37–59

Green DN (2013a) Nuu-chah-nulth First Nations' *Huulthin* (Shawls): historical and contemporary practices/Stella Blum grant report. Dress J Costume Soc America 39(2): 153–201

Green DN (2013b) Mamuu: to weave/to work. Documentary film (22 minutes). Ethnographic Film Unit at the University of British Columbia, Vancouver

Green DN, Chuuchkamalthnii (2010) Histakshitl Ts'awaatskwii (we come from one root). Documentary film (67 minutes). Ethnographic Film Unit at the University of British Columbia, Vancouver, BC

Harris C (2002) Making native space: colonialism, resistance, and reserves. University of British Columbia Press, Vancouver

Harris DC (2008) Landing native fisheries: Indian reserves and fishing rights in British Columbia. UBC Press, Vancouver

Holm B (1965) Northwest Coast Indian art: an analysis of form. Douglas & McIntyre, Vancouver

Hoover AL (ed) (2002) Nuu-chah-nulth voices: histories, objects, and journeys. Royal British Columbia Museum, Victoria

Jensen D, Sargent P (eds) (1987) Robes of power: totem poles on cloth. University of British Columbia Press in Association with the UBC Museum of Anthropology, Vancouver

Kaiser SB (2001) Minding appearances: style, truth, and subjectivity. In: Entwistle J, Wilson E (eds) Body dressing. Berg, Oxford

Kaiser SB (2012) Fashion and cultural studies. Berg, London

Kaiser SB, Nagasawa RH, Hutton SS (1995) Construction of an SI theory of fashion: Part 1. Ambivalence and change. Cloth Text Res J 13(3):172–183

Kenyon S (1980) The Kyuquot way: a study of a West Coast (Nootkan) community. Canadian Ethnology Service Paper 61, Mercury Series. Canadian Museum of Civilization, Hull

Knight R (1996) Indians at work: an informal history of native labour in British Columbia, 1848–1930. New Star Books, Vancouver

Lillethun A, Welters L, Eicher JB (2012) (Re)Defining fashion. Dress J Costume Soc America 77–99

Lutz JS (2008) Makúk: a new history of aboriginal-white relations. University of British Columbia Press, Vancouver

Massey D (1994) Space, place and gender. Polity Press, London

Massey D (2005) For space. Sage, London

Menzies CR, Butler CF (2008) The indigenous foundation of the resource economy of British Columbia's North Coast. Labour/Le Travail 61(Spring):131–149

Merleau-Ponty M (1962) Phenomenology of perception. Routledge, London

Richardson M (1982) Being-in-the-market versus being-in-the-plaza: material culture and the construction of social reality in Spanish America. American Ethnologist 9(2):421–436

Roth S (2012) Culturally modified capitalism: the native Northwest Coast artware industry. Doctoral Dissertation, University of British Columbia, Vancouver

Roth S (2015) Northwest Coast artware: beyond the use of indigenous images as 'Logos'. In: Proceedings of the virtuous circle summer cumulus conference, Milan, Italy

Sapir E (n.d.) Unpublished field notes. American Council of Learned Societies Committee on Native American Languages. Manuscript Collection, Mss.497.3.B63.c. American Philosophical Society Archives, Philadelphia

Sapir E, Swadesh M (1955) Native accounts of Nootka ethnography. Indiana University, Research Center in Anthropology, Folklore, and Linguistics, Bloomington

Sapir E, Swadesh M, Linguistic Society of America (1939) Nootka texts; tales and ethnological narratives, with grammatical notes and lexical materials. Linguistic society of America, University of Pennsylvania, Philadelphia

Sproat GM (1868) Scenes and studies of savage life. Smith, Elder and Co, London

Swan J (1868) The Indians of cape flattery, at the entrance to the straight of Fuca, Washington territory. Smithsonian Contributions to Knowledge, Washington, DC

Tarlo E (1996) Clothing matters: dress and identity in India. University of Chicago Press, Chicago

Taussig MT (1986) Shamanism, colonialism, and the wild man: a study in terror and healing. University of Chicago Press, Chicago

Thompson C (n.d.) TlaKwa designs. http://www.tlakwadesigns.com. Date visited: 15 Oct 2015

Turner T (1980) The social skin. In: Cherfas J, Lewin R (eds) Not work alone: a cross-cultural view of activities superfluous to survival. Temple Smith, London

Wilson E (2003) Adorned in dreams: fashion and modernity. Rutgers University Press, New Brunswick

Wolf ER (1982) Europe and the people without history. University of California Press, Berkeley

Korean Traditional Fashion Inspires the Global Runway

Kyung Eun Lee

Abstract Currently, a growing number of luxury fashion brands present their pre-season collections adopting Korean aboriginal fashion in textile and apparel designs. This is mainly because of the significance of South Korea as the third largest luxury market in Asia. Traditional Korean design and techniques are eco-friendly in their use of materials, processes, and design techniques. However, the sustainability of Korean aboriginal fashion has not been recognized in the global market. Global luxury brands' recent adoption of Korean aboriginal inspirations can certainly help to create global awareness of Korean aboriginal fashion's sustainability and aesthetics. The consumer trends seeking sustainable products and manufacturers' enhanced production capabilities generate new opportunities for Korean fashion brands in the global market. The present case study about Korean aboriginal fashion helps to clarify the growth of Korean aboriginal fashion in the global fashion industry. This research addresses how Korean aboriginal fashion can affect sustainable fashion consumption of global consumers. Specifically, this case study has explored the adoption of Korean traditional natural materials, dyeing techniques, and design technologies as reflected in the global fashion brands' runway collections.

Keywords Korean aboriginal fashion · Natural dyeing · Natural fabric · Sustainability

K.E. Lee (✉)
Department of Apparel, Events and Hospitality Management, Iowa State University, 28 Mackay Hall, Ames, IA 50011, USA
e-mail: Kyungeun@iastate.edu

1 Introduction

1.1 Background

For today's overheating globalization of the fashion market, fashion designers struggle to differentiate their products from global competitors in every possible way. Aboriginal fashion styles, especially those from Asia, have become popular inspirations for fashion designers who want to create unique collections to be presented on the global runways. For instance, since the mid-twenty-first century there have been growing numbers of luxury fashion houses that present the exotic pre-season (cruise) collections inspired by traditional Asian fashion styles (Ruth Styles 2015). These luxury collections reflect these styles for three major reasons: (1) to create niche market using its cross-cultural design linkage in the global fashion industry; (2) to build buzz for Asian consumers; (3) to design for jetsetters who like to travel overseas (Fernandez 2015; Maynard 2000; Ruth Styles 2015).

South Korea has been one of the major target countries in which global luxury brands adapt South Korean traditional fashion styles, materials, and design techniques to their runway collections. This is mainly because of the significance of South Korea as the third largest luxury market in Asia (Ruth Styles 2015). Chanel was the first luxury brand to present the cruise collection inspired by *Hanbok*, the traditional Korean costume, in Seoul, Korea, in 2015. The collection includes the apparel designs that reinterpret Korean traditional natural materials, patterns, and motifs. Their apparel was made with natural fabrics such as silk, hemp, and cotton, which were developed using authentic Korean textile manufacturing processes (Anderson 2015). After Chanel, other luxury brands, mostly European, such as Burberry, Louis Vuitton, and Gucci, have also been presenting Korean aboriginal styles in their pre-season (cruise) collections (Fernandez 2015). As the cases demonstrated in these collections of luxury brands, Korean aboriginal fashion has its identity in its novel textiles, natural fabrics, dyeing methods, and design techniques. These textiles are not only aesthetic for having excellent color properties and unique textures, but also eco-friendly, as the materials used and manufacturing processes generate almost no harmful environmental impacts (Troa 2015).

Therefore, adoption of Korean aboriginal fashion styles can provide multiple benefits to fashion brands: (1) promotion of sustainable fashion business practices; (2) enhancement of product aesthetics integrating cross-cultural design elements; (3) creation of a new niche market in target Asian countries which admire Korean fashion. Fashion brands can use these benefits to differentiate their brand images and products in the global market with unique and practical approaches.

1.2 The Purpose of Study

A review of the literature demonstrates that there is a lack of research on Korean aboriginal fashion in the global market. Despite some Korean fashion brands continuously investing their efforts to promote the advantages of Korean aboriginal fashion to global consumers, sustainable aspects of Korean aboriginal fashion have limited acknowledgment. Therefore, the purpose of this case study was to investigate Korean aboriginal fashion's contributions as a facilitator to sustainable fashion product consumption. More specifically, the objectives of this research were: (1) to investigate sustainability of Korean aboriginal fashion focused on the use of natural materials and manufacturing processes; (2) to determine strategic use of the global runways and exhibitions to promote Korean aboriginal fashion; (3) to discover limitations and solutions to increase adoption of Korean aboriginal fashion by global fashion designers; (4) to forecast future opportunities and challenges of Korean aboriginal fashion in the global fashion industry.

1.3 Methodology

This study adopts a qualitative approach using a review of secondary data provided by information sources in Korea. The invitations to become a part of the study were distributed to potential participants involved in Korean traditional natural materials and dyeing methods and design techniques in September 2015. The respondents who agreed to participate in this case study were requested to provide materials that explain their business activities related to Korean aboriginal fashion. The eight selected participants provided documentations for this study: (1) the branded fashion companies (Isae Fnc, Troa, Lee Young Hee, and Lie Sang Bong); (2) art institutions (BLANK SPACE and Natural Dying Culture Center); (3) Mr. Jung Kwan Chae, an artisan and the 115th Important Intangible Cultural Asset of Korea who specializes in traditional Korean natural dyeing methods.

1.4 Significance of the Study

The present case study of Korean aboriginal fashion contributes to increasing awareness of the advantages of Korean aboriginal fashion (e.g., sustainable material and process usage, aesthetic enhancement, new niche market opportunities) and its adoption to the global luxury brands. This research helps to promote Korean aboriginal fashion to global consumers by suggesting the strategic use of the global runways. Designers and company owners can enhance both sustainability and aesthetics of their fashion products by referring to this study of natural materials and process usage of traditional Korean textiles. This study is a case example of aboriginal fashion that inspires global runways to adopt sustainable fashion practices in glamourous ways. Ultimately, this study facilitates eco-design and sustainable production and consumption of fashion products in the global fashion industry.

2 Aboriginal Fashion

2.1 Definition of Aboriginal Fashion

Aboriginal means the ultimate preservation of the primeval or the exotic, which also means traditional (Geczy 2013). The term "aboriginal" originated from a European construct; aboriginal art, however, refers to a multinational cultural basis, created in different political and geographical, as well as urban and suburban, environments (Maynard 2000). Conventional aboriginal art emerged in the 1970s and became popular by the 1990s (Geczy 2013). In fashion, the significant use of aboriginal inspirations was made by fashion designers who wanted to refresh common European moods in their clothing designs. For instance, in the early twentieth century, European designers, such as Poiret, started to use aboriginal motifs and cultural inspirations in their collections (Maynard 2000). In the 1970s and 1980s, Australian fashion designers, such as Linda Jackson and Jenny Kee, aggressively accepted Australian aboriginal fashion styles instead of European inspirations, which had dominated the fashion trends in Australia for a long time (Maynard 2000; Schapiro 1980). At the time that aboriginal inspiration became popular, both the fine art and textile fields were affected by cross-cultural application at various degrees of native fashion: high, fine, and low (Eller 1997; Geczy 2013). Hence, such influences cannot be determined in uniformed styles or regulated aesthetic hierarchies (Lipovetsky 1994).

Ever since globalization and its impact on the fashion industry became more serious, aboriginal fashion inspirations have been used as mediums to differentiate the fashion brands in the light of common western concepts of fashion styles (Eller 1997; Geczy 2013). In today's fashion industry, aboriginal inspirations allow fashion designers to weave design ideas in cross-cultural approaches between typical western fashion cores and native arts and designs, so they provide timeless and authentic values in clothing, which are distinctive from competitors' products (Geczy 2013).

2.2 Aboriginal Fashion Style

In dress, the term, "style" normally refers to the identifiable and visible qualities that are connected to self, social, cultural, and political processes. There are countless variations of aboriginal arts and designs that use ancient motifs and cultural reflections in each society or through the exchange of different societies around the globe (See 1986). Therefore, aboriginal fashion styles cannot be separated from national identities. Western and non-western styles of dress interact and constantly evolve in dynamic correlations with each other, based on national identities (Childs and Williams 1997; Craik 1994). Since the 1970s, "style" has embraced more complex meanings and extensive cultural contexts (Evans 1997). Evans (1997) suggested the transformability of styles and identities related to cultures and subcultures that continuously move and change to become new style substances. In today's fashion industry, aboriginal styles in dress have diversely evolved, so they can neither be simply defined by specific forms of wearing nor be differentiated between authentic and inauthentic styles (Childs and Williams 1997; Geczy 2013).

2.3 Korean Aboriginal Fashion

There is a lack of awareness of the advantages of Korean aboriginal fashion in the global market. Sustainable aspects of Korean aboriginal fashion especially have had a limited evaluation in the literature, whilst its aesthetics have been discussed more widely. To explore these advantages without omitting essential components, the history, national identity, and types of Korean aboriginal fashion must be explored simultaneously because of their associated social, political, and cultural complexity.

2.3.1 Historical Footnotes of Korean Aboriginal Fashion

Because Korean traditional dress has over 5,000 years of history, it forms a significant part of Korean cultural heritage (Suk 1974). In the mid-nineteenth century, the cultural revolution of Korea, "*Gabo* Reforms," promoted the modernization of both Korean society and dress. As the social class systems collapsed after the reform, Korean dress was rapidly westernized when the types of clothing did not represent social classes of people. Until the 1940s, Korean fashion had a strong tendency to follow western dress styles and trends. However, beginning in the 1950s, fashion designers started to use Korean aboriginal inspirations in their collections. The 1960s and 1980s were when Korean fashion gradually industrialized and flourished. In the 1970s, the changes and developments in Korean fashion consisted of both western dress influences and Korean aboriginal inspirations (Germ et al. 1992). During this period, fashion designers constructed the collections using Korean traditional materials, colors, patterns, and design techniques from the eighteenth and nineteenth century trends (Germ et al. 1992). With continued advancement of the Korean fashion industry until the 1990s, Korean fashion products started being exported to overseas market. Among exported goods, those with Korean aboriginal inspirations, such as traditional motifs and design techniques, were popular in the global fashion capitals including Paris and New York (Kim 2012a, b). In the 2000s, Korean students who graduated from advanced

Fig. 1 Designer Lie Sang Bong's New York Fashion Week, spring-summer collection entitled "Dream Road" inspired by the beauty of nature, consisting of butterflies, flowers, and clouds (LIE SANGBONG 2015)

fashion schools in foreign countries, such as Italy, France, and the U.S., played important roles in transforming Korean aboriginal fashion styles into distinctive cross-cultural design languages for global consumers.

Currently, some early pioneers and new generations among Korean fashion designers collaborate to develop innovative design ideas to promote Korean fashion in the global market (Kim 2012a b). For instance, in 2012, the Council of Fashion Designers in Korea (CFDK) was established, and the president of CFDK is the world preeminent designer Lie Sang Bong. He has dedicated his entire career to promoting the high profile global presence of Korean fashion through reflecting strong Korean aboriginal fashion heritage in his designs (Kim 2015). The significant influences on the appearances of traditional Korean dress were made by social classification, self-identity, and counterculture throughout Korean history (Higgins and Eicher 1995; Yoon and Yim 2015). Currently in Korea, self-identity expressed by modern western clothing is popular, because of the collective culture and value of Korean society in which western culture and lifestyles are considered to be trendier and stylish (Yoon and Yim 2015); see Fig. 1.

2.3.2 National Identity of Korean Aboriginal Fashion

The national identity of Korean aboriginal fashion is formulated by the strong forces of tradition, nature, and culture (Germ et al. 1992). These forces are critical to understanding the evolutions of Korean fashion (Germ et al. 1992). The national identity of Korean traditional dress embraces a strong historical heritage, which provides cross-cultural linkages in design between global fashion designers and Korean aboriginal art and textiles. The national identity of Korean aboriginal fashion can be attributed to three traditional cultural assets: (1) *Hanbok* (costume); (2) symbols and motifs; (3) *Obangsaek* (colors).

Hanbok

Because the Korean traditional dress originated in the Goguryeo Kingdom (37 BC–668 AD), it is called *Hanbok*, which used to be everyday clothing for both men and women (Korea.net 2014). In today's Korean society, *Hanbok* is worn for special occasions, such as weddings and receptions, more often by women than men. However, it still remains an important part of the Korean cultural heritage (Germ et al. 1992). Roach and Musa (1980) explained that traditional dress is grounded in a dominant culture of a society and represents a certain meaning, whereas fashionable dress commonly implies a particular segment of the culture within a specific time frame. In the case of *Hanbok*, it contains the combined meaning of both traditional and fashionable dress, whilst these two forms affect each other (Flugel 1969).

A general composition of *Hanbok* includes a jacket, vest, and outerwear, and adds a skirt for women and pants for men (Germ et al. 1992; Korea.net 2014). The basic shape of *Hanbok is* curved and volumized. The jacket *Chogore,* specifically, has a dramatically curved hemline and underarm (Germ et al. 1992). The multiple undergarment layering is to create fuller silhouettes and it is one of the most distinctive characteristics of traditional Korean dress (Kim 2012a, b). There are several major ornaments and accessories of *Hanbok*: *Dong-Jung* (the white paper

Fig. 2 Designer Lee Young Hee's *Hanbok* collection presented in 2013 fashion show in Seoul, Korea (Lee Young Hee 2015)

neck band), *Norigae* (the hanging ornament), and *Buson* (the padded socks) (Germ et al. 1992; Korea.net 2014). The common materials used to make *Hanbok* are traditional natural fabrics such as silk, hemp, linen, and cotton (Germ et al. 1992: Kim 2012a, b). Following the evolution in Korean fashion history, *Hanbok* has been constantly changing in different major types of designs (Flugel 1969; Germ et al. 1992); see Fig. 2.

Symbols

Korean aboriginal inspirations include symbols created by the strong shamanism influence which prevailed in the ancient Korean society (e.g., plants, flowers, birds, animals, rock, water, cloud) (Grayson 1992). There are two major shamanistic influences of symbols in traditional Korean dress: (1) nature; (2) *Yin* and *Yang*. Most of the symbols used in *Hanbok* design are reflected in nature as originating in the agricultural-based ancient Korea, and they were the visual methods of expressing people's appreciation towards nature (Korea.net 2014). Other frequently used symbols are those related to *Yin* and *Yang,* meaning the harmony of nature (Kim 2006a, b). For instance, the contrasting colors of blue and red illustrate the meaning of *Yin* and *Yang*, which represent the harmony of men and women, moon and sun, and shadow and light (Kim 2006a, b).

Colors (Obangsaek)

Obangsaek is based on *Yin* and *Yang* and it refers to the combination of five directions and colors—north (black), south (red), east (blue), west (white), and center (yellow) (Kim 2006a, b). *Obangsaek* describes the order of nature in the harmonized universe (Germ et al. 1992; Kim 2006a, b). In addition to white, these five colors of *Obangsaek* are the most commonly used colorways of *Hanbok* (Kim 2006a, b). In *Hanbok,* the nature symbols and *Obangsaek* are reflected in the motifs, printing patterns, and embroideries for surface design of textiles (Germ et al. 1992). The social classes of ancient Korean society were described in clothing by the use of different combinations of these traditional symbols and colors (Germ et al. 1992). Compared to traditional styles, contemporary *Hanbok* designs still reflect bold Korean heritage. However, the meaning of the symbols and colors are not strictly applied (Germ et al. 1992; Kim 2006a, b).

There were two global showcases for *Obangsaek* presented in the global fashion industry, which were the New York Fashion Week and Les Arts Décoratifs Exhibition. In 2012, Concept Korea, a part of New York Fashion Week, presented a special project featured at *Obangsaek* at Lincoln Center in New York to generate global awareness of Korean aboriginal fashion (Kim 2012a, b). The project reflected Obangsaek themes, which were composed of a modern dance performance made by Martha Graham Dance Company and five Korean designer brands (Lie Sang Bong, Choiboko, Cres. E Dim., Kye, and Son Jung Wan) were also featured in the Spring/Summer 2013 collection runway shows (Korea.net 2012). This event was a great success and had drawn global attention from media, journalists, and influential fashion ambassadors such as Fern Mallis (New York Fashion Week founder), Philip Bloch (Vogue stylist), and Colleen Sherin (fashion director for Saks 5th Avenue department store) (Kim 2012a, b; Korea.net 2012).

At the 2015 Les Arts Decoratifs (The Paris Museum of Arts and Crafts), designer Lee Young Hee exhibited her Hanbok collection reinterpreted Obangsaek in contemporary tastes along with other world famous designers including Karl Lagerfeld (Park 2015). This exhibition was to celebrate the 130th anniversary of Korea–France diplomatic relations and drew approximately 15,000 visitors (Park 2015).

Hangul

Hangul is the Korean alphabet which was invented in the mid-fifteenth century (The Economist 2013). What makes *Hangul* special is the fact that it was created from scratch in accordance with the Korean language, which was not evolved from pictographs or other emulated lettering systems (The Economist 2013). The Korean Wave is the phenomenon of Korean pop-culture (e.g., music, movies, TV drama) global popularity growth (Lee 2014a, b, c). It has made Korean culture popular on a global level. This has played a vital role in the growing interest and global trend among people's desire to learn the Korean language and Hangul (Lee 2014a, b, c). Hence, *Hangul* has become effective for design patterns and motifs to promote Korean aboriginal fashion in the global market.

Currently, many Korean designers and artists are featuring *Hangul* in their creative work. For instance, in 2005, designer Lie Sang Bong's *Hangul* motif designs were first introduced to global audiences on the Paris prêt-a-porte exhibition and showcased in the 2006 Paris Fashion Week, Fall-Winter 2006 prêt-a-porte Collection, and Fall-Winter collection (LIE SANGBONG 2015). Since then, Lie Sang Bong has been continuously showcasing unique *Hangul*-inspired designs in his global runway collections (LIE SANGBONG 2015). For these collections, Lie Sang Bong reinterprets the traditional *Hangul* motif into wearable art, which incorporates contemporary design technologies such as digital printing and illustration beyond traditional black and white calligraphic designs (LIE SANGBONG 2015); see Fig. 3.

In 2012, a famous Korean actor, Yoo Ah-In, launched his Hangul themed T-shirt collection in collaboration with the Korean fashion brand Nohant (Lee 2014a, b, c). This collection was promoted through the Korean TV drama "Fashion King," in which he acted as a fashion designer (Lee 2014a, b, c). These two cases of LIE SANGBONG and Yoo Ah-In are the novel examples of naturalizing a Korean cultural heritage component, *Hangul,* as a marketable pop-culture in innovative design approaches.

Fig. 3 Designer Lie Sang Bong's *Hangul* collection presented in Paris Fashion Week, Spring-Summer 2007 prêt-a-porte Collection (LIE SANGBONG 2015)

2.4 Korean Aboriginal Fashion and Sustainability

Focused on environmental impacts of clothing, there are three distinctive forms of aboriginal arts and designs in traditional Korean dress: (1) using natural materials; (2) adopting natural dyeing methods; (3) applying design techniques.

2.4.1 Using Natural Materials

In Korean traditional dress, there have been many different natural materials developed throughout its history. Among different traditional Korean natural fabrics, those often used by today's fashion designers include the following: (1) *Hanji-Sa* (paper mulberry); (2) *Hansan-Mosi* (hemp); (3) *Moo-Myung* (cotton); (4) *Myung-Joo* (silk); (5) *Chun-Po* (union cloth of silk and hemp); (6) *Sam-Be* (ramie).

Hanji Fabric (Paper Mulberry)

The term *Hanji* means Korean paper, made with the inner bark materials of the paper mulberry tree and originating over 1,600 years ago (Choi et al. 2012; Kim 2006a, b). The paper mulberry tree is one of the most eco-friendly materials, because of its rapid growth and high germinative power (Xu et al. 2011). Durability of *Hanji* was illustrated by the excellent condition of nine *Hanji* publications from the fourteenth to early twentieth centuries, which were listed in the Memory of the World Register by UNESCO (United Nations Educational, Scientific, and Cultural Organization) (UNESCO n.d.). *Hanji-Sa* is a yarn made from *Hanji* paper, which is manufactured using adapted Korean traditional paper making techniques, molding, layering, and burnishing, to produce strength and flexibility (Choi et al. 2012; Kim 2006a, b). Finished *Hanji* paper is converted into yarn through the processes of slitting, twisting, and weaving (Isae 2015; Troaco.com n.d.). These processes have minimal negative environmental impact by using natural materials and generating less toxic water waste (Troaco.com n.d.).

Since the late twenty-first century, Korean textile manufacturers have started to develop *Hanji* fabric to be consumed in Korean and overseas markets (E-daily News 2013). The novel fiber structures created through paper to yarn transformation processes allow *Hanji* fabric to possess advanced technical performance-enhancing wearing comfort of clothing (e.g., breathability, wet-strength, ventilation, moisture absorption, antimicrobial properties) (Choi et al. 2012; Jeong et al. 2014; Kim 2006a, b). *Hanji-Sa* is flexible, allowing for both weaving and knitting from low to high yarn counts, depending on the purpose of the clothing (Jeong et al. 2014). Hanji clothing designs have been featured by many Korean designers on global runways to promote eco-friendliness and other benefits of the fabrics. For instance, designer Lie Sang Bong featured the *Hanji* lingerie collection in collaboration with the Korean lingerie company Cowell fashion in 2011 (Jang 2011). To enhance aesthetics of this collection, Lie Sang Bong used digital printed calligraphic *Hangul* design, the Korean alphabet blending black and white colorways (Jang 2011). Since 2004, designer Han Song, of the fashion brand Troa, has been developing contemporary textiles and apparel made with *Hanji* fabrics (Troaco.com n.d.). Song currently sells the extensive *Hanji* denim collections, made using natural dyeing methods, to global trend-setting retailers such as Barneys New York, Harvey Nichols, Ten Corso Como, and Collette (Troa 2015). Both designers, Lie Sang Bong and Han Song, made distinctive examples

Fig. 4 Troa's *Hanji* denim collection naturally dyed with Korean ink, natural indigo, and soapberry tree (Troa 2015)

of expanding the use of traditional Korean fabric *Hanji-Sa* to the broader product types, like lingerie and denim; see Fig. 4.

Hansan-Mosi (Ramie)

Hansan-Mosi is a plain-woven fabric made with the peel of the ramie plant and has been used as an ideal summer textile for over 1,500 years in Korea (Kim 2012a, b). *Hansan* is the name of the *Mosi* manufacturing hub area in the *Secheon-Gun* region in the *Choongnam* province, Korea, in which the ideal natural conditions are available for cultivating high quality ramie plants without using a large amount of fertilizers (Cultural Heritage Administration n.d.b; Kim 2012a, b). Along with the fine qualities and soft textures of the fabrics, *Hansan-Mosi* has multiple technical advantages for summer clothing such as breathability, moisture absorption, fabric strength, and anti-yellowing color properties without artificial finishing (Cultural Heritage Administration n.d.b). Together with *Moo-Myung* (cotton), *Hansan-Mosi* is one of the fabrics most threatened by mass produced textiles in the modern fashion industry (Kim 2012a, b). Despite the rapid collapse of the old Korean weaving culture in general, traditional weaving techniques are still widely used to manufacture *Hansan-Mosi* in Korea (Kim 2012a, b). Because *Hansan-Mosi* was listed in the Memory of the World Register in 2011 by UNESCO, it receives global attention (Cultural Heritage Administration n.d.b). Hence, *Hansan-Mosi* is in a more advantageous position to preserve its traditional weaving culture and techniques, compared to other traditional Korean fabrics (Korea.net n.d.a).

Hansan-Mosi performs significant roles in promoting Korean traditional fashion to global audiences through Korean fashion designers' global exhibitions and runway shows. For instance, in the 2011 Paris Haute Couture Fashion Week, Lee Young Hee, the *Hanbok* designer, showcased her *Hansan-Mosi* collection, which was reconstructed in western dress silhouettes (Park 2015); see Fig. 5.

In 2014, at White, the global trade show during Milano Fashion Week, Isae presented the hand-made wearable art collection, composed of a total of 30 pieces

Fig. 5 Designer Lee Young Hee's *Hansan-Mosi* collection presented in 2011 Paris Haute Couture Fashion Week Lee Young Hee (2015)

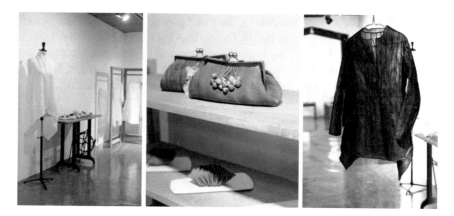

Fig. 6 Isae's *Hansan-Mosi* collection presented in white during 2014 Milano Fashion Week (Isae 2015)

in apparel, bags, accessories, and home decor crafts, created in collaboration with several Korean textile artists (Ahn 2014); see Fig. 6.

Moo-Myung (Cotton)

In Korean clothing history, a plain-woven cotton fabric, *Moo-Myung*, was one of the most commonly used materials, and was worn by commoners rather than upper class people (The Academy of Korean Studies n.d.). Because of *Moo-Myung*'s symbolic image of white, Korean people are called "the white-clad folk" (Lee 2014a, b, c). The records of using *Moo-Myung* fabrics exist from the period of three kingdoms (fourth to seventh centuries). In the mid-fourteenth century, during the late Goryeo dynasty, massive *Moo-Myung* cultivation was started with the initial import of cotton seed from the Yuan dynasty of China (Shim and Park 2003; The Academy of Korean Studies n.d.). Since then, *Moo-Myung*-related textile businesses flourished until the nineteenth century (Shim and Park 2003).

The use of traditional handloom manufacturing techniques distinguishes *Moo-Myung* from western cotton fabrics, such as muslin and calico, because of its uniquely crafted fabric texture (Cultural Heritage Administration n.d.f). In addition to sun-bleach white as the major colorway of *Moo-Myung* fabrics, different colorways were also available by using natural herbal dyeing methods in the history of traditional Korean dress (Shim and Park 2003). In the ancient Korean society, *Moo-Myung* was a four-season fabric for commoners' everyday clothing and was adapted by using textile design manipulation; quilted for winter, double layered for spring and fall, and single layered for summer (The Academy of Korean Studies n.d.). It was also widely used for home textile products such as bedclothes, towels, and covers (The Academy of Korean Studies n.d.). In the late nineteenth century, after the time of civilization and enlightenment in Korea, the use of

Moo-Myung was declined, because of the rapid growth of mass-produced cotton textile products (Shim and Park 2003). To continue the tradition of *Moo-Myung*, in 1969 the handloom *Moo-Myung* weaving method called *Naju Saekgolnayee* was designated as the twenty-eighth Important Intangible Cultural Asset of Naju city in the *Jeonnam* province, Korea (Cultural Heritage Administration n.d.f).

Myung-Ju (Silk)

The 87th Important Intangible Cultural Asset of Korea *Myung-Ju*, also called *Bi-Dan*, is a luxury plain-woven silk fabric made with thread obtained by unraveling cocoons (Cultural Heritage Administration n.d.g). Among different Korean traditional fabrics, *Myung-Ju* is the only textile created from non-herbal resources (Han 2013). In the Chosun dynasty (fourteenth to twentieth centuries), high quality *Myung-Ju* started being produced and worn mostly by high class people as a symbol of social prestige (Kim 2012a, b). In today's textile industry, various *Myung-Ju* fabrics with different degrees of delicate luster and fabric softness have been developed depending on weaving methods such as silk kokura, silk bengaline, and silk voile (Doopedia n.d.c). Weight and handfeel of *Myung-Ju* can be modified by removing raw material impurity Sericine; removing more Sericine decreases the weight and stiffness of the fabric (Doopedia n.d.c); see Fig. 7.

Fig. 7 Designer Lee Young Hee's *Hanbok* collection, made with *Myung-Joo* (Lee Young Hee 2015)

Fig. 8 Isae's *Chun-Po* jacket presented in spring-summer 2011 Seoul Fashion Week (Isae 2015)

Chun-Po (Union Cloth of Silk and Ramie)

Chun-Po, also called *Chungyang Chun-Po,* is a union cloth that is made up of *Myung-Joo* (silk) warp and *Mosi* (ramie) weft yarns to generate unique appearances and handfeel (Cultural Heritage Administration n.d.a; Kim 2012a, b). It emerged in the late *Cho-Sun* dynasty around the eighteenth century and was popularly used as luxury clothing textiles in the 1940s (Cultural Heritage Administration n.d.a). Even though the fabric texture of *Chun-Po* is similar to *Hansan-Mosi*, it is differentiated in aesthetics by its harmony of stiffness and pliability (Isae Blog 2011; Kim 2012a, b). Distinctive textile performances that integrate insulation of silk and cooling sensation of ramie improve the advantages of clothing for transitional seasons in early spring and summer (Cultural Heritage Administration n.d.a; Isae Blog 2011). A major development of *Chun-Po* has been made in the *Chungyang-Gun* region in the *Choongnam* province, Korea (Cultural Heritage Administration n.d.a).

In today's textile industry, the popularity of *Chun-Po* has decreased, because of high prices for handcraft manufacturing and consumer preference for synthetic fabrics (Doopedia n.d.b). As a result, there are currently limited numbers of textile experts who inherited *Chun-Po* weaving techniques mainly from *Chungyang-Gun* (Doopedia, n.d.b). To maintain the existence of *Chun-Po*, it was registered as the 25th Important Intangible Cultural Asset of the *Choongnam* province, Korea in 1998 (Cultural Heritage Administration n.d.a; Isae Blog 2011). Some branded fashion companies in Korea use *Chun-Po* to assist their premium product lines in promoting the fabric's advantages. For instance, Isae, the sustainable Korean fashion brand, launched a jacket made with *Chun-Po* using natural indigo and persimmon dyeing methods in 2011 (Isae Blog 2011). This jacket was distributed to the company's flagship stores and major prestige department stores and contributed to consumer awareness of *Chun-Po* through social media and in-store promotions (Isae Blog 2011); see Fig. 8.

Sam-Be (Hemp)

Plain-woven fabric *Sam-Be* is made with hemp bast fiber, providing multiple benefits to be used for summer clothing (e.g., moisture absorption, ventilation, heat conduction, abrasion resistance, moisture strength) (Doopedia n.d.a). *Andong-Po* is high quality *Sam-Be* produced in Andong city of the Youngam province, Korea, where the most favorable climate and soil conditions are available for hemp cultivation. In 1970, *Andong-Po* was registered as the first Intangible Cultural Asset of the Andong region in the Gyeongbuk province, Korea (Cultural Heritage Administration n.d.c). There are two types of *Sam-Be*, *Saengnaengi* and *Iknaengi*; *Saengnaengi* is woven with non-steamed hemp, which was previously consumed by the prestige class, and *Iknaengi* uses steamed hemp worn by commoners (Bae 2003). *Sam-Be*'s sustainable and elaborate general making processes involve the following: (1) stripping off harvested hemp leaves; (2) steaming and peeling off the hemp sheaf; (3) sun-drying; (4) splitting the hemp along the grain to make yarn; (5) connecting and starching yarns; (6) handloom weaving (Cultural Heritage Administration n.d.c; Kim 2012a, b). The Goryeo dynasty (tenth to fourteenth centuries) was the most flourishing period for *Sam-Be* development. However, in the Chosun dynasty (fourteenth to twentieth centuries), the popularity of *Moo-Myung* (cotton) cultivation decreased the growth of *Sam-Be* textile businesses (Cultural Heritage Administration n.d.c).

2.4.2 Adopting Natural Dyeing Methods

Natural Dyeing

There are records of traditional Korean natural dyeing identified from the Goguryeo Kingdom (37 BC–668 AD) (Natural Dyeing Culture Center 2007a). In some agriculture guidebooks published in the nineteenth century, different dyeing materials, methods, and mordants were introduced (Natural Dyeing Culture Center 2007a, b, c, d). Traditional Korean natural dyeing techniques are eco-friendly for three major reasons: (1) using natural materials for both dyes and mordant, which are biodegradable and recyclable; (2) generating almost no harmful water-waste throughout the dyeing processes; (3) dyeing materials have strong viability in bad environment conditions for growing (Isae 2015). There are different categories of traditional Korean natural dyeing materials, depending on the biological (e.g., vegetable, animal, mineral) and chromatic characteristics (e.g., monochromatic, polygenetic, natural pigment) (Natural Dyeing Culture Center 2007b). For instance, indigo and safflower are monochromatic vegetable dyes. Red clay and mud are mineral natural pigment dyes (Natural Dyeing Culture Center 2007b). In today's fashion industry, there are five natural dyeing materials frequently used: indigo, Korean ink, mud, persimmon, and safflower seed (Natural Dyeing Culture Center 2007b). Among these materials, indigo involves the most difficult and complicated method, requiring elaborate manufacturing techniques (Isae 2015). In addition to natural dyeing materials, natural mordant dyes (e.g., lye, alum, salt) are a significant part of traditional Korean dyeing methods, because of their contribution to color fixation and affinity in sustainable approaches (Natural Dyeing Culture Center 2007b).

Yeomsaekjang refers to a master craftsman who specializes in dyeing fabrics using natural materials that contain herbal, mineral, and animal resources, and such an occupation existed in the royal court of the Chosun dynasty (fourteenth to twentieth centuries) (Cultural Heritage Administration n.d.d). The occupational technique was registered as the 115th Intangible Cultural Asset of Naju city of the Jeonnam province, Korea in 2001 (Cultural Heritage Administration n.d.d). To ensure the continued success of traditional Korean natural dyeing culture and techniques, designated artisans of *Yeomsaekjang* have committed their lives to transferring Korean natural dyeing culture and techniques through exhibitions, seminars, and training classes (Jung 2015). Together with the efforts made by artisans, some Korean fashion companies, such as Isae and Troa, have been investing in research and development (R&D) to develop contemporary natural dyeing methods adapted from traditional Korean techniques, with the aim of enhancing production efficiencies and dyeing qualities (Isae 2015; Troa 2015).

Natural Indigo Dyeing

Traditional Korean indigo dyeing was initiated in the Joseon dynasty (fourteenth to twentieth centuries) and it was continued through the 1950s (Jung 2015). After inactive periods caused by the Korean War (1950–1953), traditional indigo dyeing was restarted in the 1970s and has been flourishing since the 1980s (Jung 2015). Polygonum, the main material for natural indigo dyeing, is eco-friendly, not only because it is an herbal dye, but also for its survival in frequent flooding along the rivers (Isae 2015). Polygonum indigo dyeing provides benefits to both technical (e.g., antimicrobial, deodorization, natural cooling, fabric strength, mothproof) and physiological (e.g., sterilizing action, anti-allergic reaction, blood circulation) performances (Isae 2015). General Korean traditional natural indigo dyeing methods are executed through the following processes: (1) harvesting the polygonum indigo plants normally from July through August; (2) putting the harvested indigo in a large water-filled vessel for 2–3 days; (3) removing the plants from the vessel; (4) mixing with the lime solution extracted from burnt oyster shells; (5) leaving the vessel until the water turns green and then indigo with bubbles; (6) collecting indigo sediment after removing impurities from the vessel; (7) boiling the collected indigo mixed with lye to allow water solubility; (8) soaking the cloth into the finished indigo-lye solution for 3–5 min; (9) air-drying; (10) washing in warm water and drying before use (Jung 2015); see Fig. 9a, b.

(a)

(b)

Fig. 9 **a** Traditional natural indigo dyeing processes presented by Jung, Kwan Chae, a designated artisan of 115th intangible cultural asset of Naju city in Jeonnam province, Korea (Jung 2015). **b** Isae's natural indigo dyeing collection presented in 2011 FW Seoul Fashion Week (Isae 2015)

Korean Ink Dyeing

Korean ink is an essential component of traditional Korean calligraphy, together with paper, brush, and ink stone (Plus Korea Times 2008). The use of Korean ink for writing was recorded in documents published during the Goguryeo Kingdom (37 BC–668 AD) (Plus Korea Times 2008). Korean ink is a natural material made with burnt pine tree roots, and it generates almost no harmful effects on the environment (Isae 2015). Traditional Korean ink dyeing processes are as follows: (1) grinding an ink stick to make ink; (2) diluting the ink with warm water; (3) soaking the fabric in the ink solution; (4) boiling the fabric with vinegar; (5) washing;

Fig. 10 Troa's Korean ink dyeing collection, made with *Hanji fabric,* presented in 2011 FW Seoul Fashion Week (Troa 2015)

(6) heating when the fabric is wrapped to create soft-gloss (Isae 2015). Troa is one of the major fashion brands that frequently uses Korean ink dyeing methods for the global runway collections. In the Fall-Winter 2012 Seoul Fashion Week, Troa presented a unique collection that combined Korean ink dyeing, *Hanji* blended fabrics, and digital printed motifs of *Minhwa*, the Korean folk art formed in the seventeenth century (Troa 2015); see Fig. 10.

Persimmon Dyeing

Conventionally, persimmon dyeing has been mainly used in the Jeju Island and seaside areas in Korea (Natural Dyeing Culture Center 2007c). One of the most distinctive advantages of persimmon dyeing is the use of condensed tannin ingredients that helps to produce stronger color properties in various brown shades (Isae 2015). Persimmon dyeing also provides benefits to the fabrics such as color fastness, antimicrobial properties, natural preservation, moisture strength, and ventilation capabilities (Natural Dyeing Culture Center 2007c). An ideal type of persimmon is a lotus-persimmon (tree), which is smaller than other persimmons (Isae 2015). Unripe persimmons, in green, generate darker colors and ripe persimmons, in reddish shades, create lighter colors (Isae 2015). Persimmon dyeing processes are relatively simple, compared to other natural herbal dyeing methods, in which the following six steps are required: (1) dividing and grinding persimmons harvested from July to October; (2) diluting persimmon paste with water (50:50); (3) juicing persimmon paste using a net sheet to extract juice; (4) adding bay salt to facilitate color formation and fabric softness; (5) steeping fabrics into persimmon juice; (6) sun-drying. These processes are normally repeated two or three times for better color properties (Isae 2015); see Fig. 11.

Fig. 11 Isae's persimmon dyeing processes (Isae 2015)

Safflower Dyeing

Since the period of the Three Kingdoms (fourth to seventh centuries), safflower has been used as one of the most dominant dyeing materials to create red shades (Natural Dyeing Culture Center 2007d). The pigments of safflower are composed of soluble yellow and insoluble red colors; hence, it is necessary to use a natural mordant such as lye (Natural Dyeing Culture Center 2007d). Traditional safflower seed dyeing methods involve the following processes: (1) harvesting safflowers approximately after 2 days from the blossoming in late June; (2) straining safflowers through a sieve to make the paste; (3) adding boiling water to the paste to remove yellow pigment; (4) removing the previous water and brewing red pigment with clean boiled water; (5) adding omija juice to acidify red pigment for color fixation; (6) soaking the fabric in the solution; (7) washing and air-drying (Jung 2015; Natural Dyeing Culture Center 2007d); see Fig. 12.

Natural DTP (Digital Textile Printing)

To increase manufacturing efficiency and generate novel dyeing effects, some textile craftsman studios in Korea developed natural digital printing methods using eco-friendly ink made from natural dyeing materials such as indigo and safflowers. For instance, Whasoomok is an R&D company specializing in natural dyeing using advanced technology in Youngchun city of the Kyungbook province, Korea (Whasoomok 2015). This R&D company developed natural DTP (digital textile printing) ink methods, which involve three main steps: (1) extracting and condensing dyeing materials using low temperature vacuum techniques; (2) powdering

| (1) | (2) | (5) | (6) |

Finished Sam-Be (ramie) fabric dyed with safflower

Fig. 12 Traditional natural safflower dyeing processes presented by Jung, Kwan Chae, a designated artisan of the 115th Intangible Cultural Asset of Naju city in Jeonnam province, Korea (Jung 2015)

using spray-drying machine; (3) mixing and boiling with a thickening agent. Low temperature vacuum techniques increase color fastness properties and dyeing affinity. Natural DTP is a unique example of natural dyeing technology incorporation, a sustainable and aesthetic product innovation approach (Hwasoomok 2015); see Fig. 13.

2.4.3 Applying Design Techniques

Nubi

Nubi is a Korean traditional quilting technique in which the outer shell and lining are sewn together, with or without inserting cotton or feather fillers, in broad stitching to construct a body of textile (Ju 2011). *Nubi* provides multiple benefits to clothing such as insulation, durability, and aesthetics. It became popular with the emergence of *Moo-Myung* (cotton) in the Chosun dynasty (Lee 2012).

(a)   **(b)**   **(c)**   **(d)**

Fig. 13 Whasooomok's natural DTP processes. **a** Spray-drying instrument. **b** Natural color sampling. **c** Color making. **d** Natural DTP sample made with acacia, nutgall, and marigold (Whasoomok 2015)

(a)

(b)

Fig. 14 **a** Isae's Nubi hand-craft making (Isae 2015). **b** Designer Lee Young Hee's *Hanbok* collection, made with Nubi technique (Lee Young Hee 2015)

The term *Nubi* has a religious symbolic meaning relating to Buddhism, as it was derived from patch details of the Buddhist monk's robe designs (Lee 2012). *Nubi* is differentiated with the application of a broad range of stitching and sewing techniques from *Chogagbo* (Bojagi), Korean patchwork wrapping cloth (Ju 2011). In accordance with the distance between stitching lines, the types of *Nubi* are divided into *Jan-Nubi* (narrow line, 0.3 cm), *Jung-Nubi* (middle spaced line, 0.5 cm), and *Dumun-Nubi* (wide line, 1.0 cm) (Ju 2011). *Nu-Bi* Jang is an expert specializing in Korean traditional *Nubi* techniques, and it was registered as the 107th Intangible Cultural Asset of Kyungju city of the Gyeongbuk province, Korea in 2001 (Cultural Heritage Administration n.d.e). Despite the high cost and long lead-time, because of the hand-craft manufacturing, the fashion brand Isae and *Hanbok* designer Lee Young Hee has been continuously investing in *Nubi* design development to protect the techniques as valuable assets of traditional Korean fashion (Isae 2015); see Fig. 14a, b.

Fig. 15 Lie Sang Bong's *Bojagi* themed collection presented in New York Fashion Week, Spring-Summer 2016 collection (LIE SANGBONG 2015)

Bojagi (Jogakbo)

Traditional wrapping cloth *Bojagi*, also called *Jogakbo*, was used by all social classes as daily necessities to cover, store, and carry objects throughout the Choson dynasty (fourteenth to fifteenth centuries) (Jeon 2011; Shin et al. 2011). The aesthetics of the traditional wrapping cloth *Bojagi* come from an incorporation of multiple design techniques such as quilting, natural dyeing, embroidery, and knotting (Jeon 2011). In traditional *Bojagi* design, different combinations of colors and patterns are distinctively decided according to the *Yin and Yang* principle praying for a long life and harmony with nature (Jeon 2011). There are two *Bojagi* types used depending on the social class of the user; *Gung-Bo* used by upper class and *Min-Bo* for commoners (Roberts and Huh 1998; Shin et al. 2011). As patchwork design is frugal to be made with small pieces of leftover fabrics, *Bojagi* was used more by commoners than upper class people (Roberts and Huh 1998). *Bojagi* is one of the most globally recognized Korean traditional design techniques used by designers in global exhibitions and fashion runway shows. For instance, designer Lie Sang Bong introduced the traditional *Bojagi* design into contemporary design using digital printing and laser cutting technologies for his collection presented in the New York Fashion Week Spring-Summer 2016 Collection (LIE SANGBONG 2015); see Fig. 15.

From a review of the literature, in Korean aboriginal fashion, there are four key aspects of eco-friendliness identified. First, both the use of material and processes generate almost no harmful impact on the environment throughout the lifecycle of the product; the natural materials used are biodegradable and recyclable and water waste generated from manufacturing processes contains almost no toxic material. Second, Korean aboriginal fashion provides multiple technical performances and benefits without using synthetic materials or processes. Third, design techniques (e.g., *Nub*i, *Bojagi*) facilitate multi-dimensional and practical use of materials such as transformation, redesign, and upcycling. Lastly, contemporary manufacturing methods which incorporate traditional materials and advanced technologies (e.g., natural digital printing, laser cutting) allow fashion designers to innovate products in unique and sustainable approaches.

3 Korean Aboriginal Fashion on Global Runways

3.1 Sustainable Movement of Global Luxury Brands

Luxury brands' involvement in sustainability is important for two main reasons: (1) heavy dependence on resources to make premium and rare products; (2) powerful influence to promote sustainability globally. Hence, in today's fashion market, many luxury brands pursue sustainable business practices through redesigning products, using eco-friendly materials and processes, and adopting green marketing approaches. For instance, since 2001, Louis Vuitton Moet Hennessy has retained an environmental stance, and the company is actively involved in environmental activities such as monitoring their carbon emission (Joy 2015). Tiffany & Co. published the corporate responsibility report in 2011 and was dedicated to responsible mining (Joy et al. 2012).

Global luxury brands have powerful influences in promoting aboriginal fashion in the global fashion market. For instance, in 2012, Marni, the Italian luxury brand, presented apparel and jewelry lines using exotic styles, mixed patterns, and prints in its global runway collections inspired by different countries' aboriginal art. Marni's jewelry was made with recycled plastic bottles and vintage records (Olsen 2011). In the same year, the collaborated collection of Marni and fast fashion brand H&M was launched; the same exotic moods and recycled materials were used, so its sustainable approaches were exposed to a wider range of consumers (Wang 2012). In addition to collaboration with Marni, H&M committed to produce their products using sustainable textiles made with recycled materials such as plastic bottles, wool, and cotton (H&M 2013). The cases of Marni and H&M are distinctive examples of a global well-known brand's power of influence to encourage the use of sustainable products inspired by aboriginal fashion in the global market.

Promoting Korean aboriginal fashion inspirations should be reinforced on the global runways for three reasons. The first is to inspire global luxury brands' adoption in their collections. The second is to facilitate global consumers' sustainable fashion product consumption in glamorous and expedited ways. The third is to create new niche-market opportunities through product innovation using aboriginal fashion as a marketable pop-culture. Ultimately, Korean apparel companies must inspire the global fashion industry to adopt Korean aboriginal fashion as an eco-friendly fashion business practice.

3.2 Korean Aboriginal Fashion on Global Runways

Throughout modern history of Korean fashion, many Korean designers have been making efforts to promote Korean aboriginal fashion in the global market. For instance, since 2002, designer Lie Sang Bong has been continuously presenting his collections in multiple global runway shows such as New York and Paris Fashion Week (Lie Sang Bong 2015). Interestingly, there are two non-Korean global preeminent designers, Caroline Herrera and Dries van Noten, who have featured Korean aboriginal fashion-themed collections on the global runways. Caroline Herrera is one of the world's leading fashion designers of the 1970s and 1980s. In 2010, her spring-summer 2011 cruise collection featured the reinterpreted Korean traditional dress *Hanbok* in New York Fashion Week (Phelps 2010). Caroline Herrera decided to create the *Hanbok* inspired collection, after she saw Korean *Hanbok* designer Lee Young Hee's collection at Concept Korea, part of New York Fashion Week that she attended (Kim 2012a, b). In Herrera's *Hanbok*-inspired cruise collection, the inspirations from a Korean traditional hat *Got*, motif, patterns, and undergarment layering were incorporated with western clothing styles (Kim 2012a, b; Phelps 2010). This was the first example where Korean aboriginal inspirations were accepted by a foreign designer to construct a new, elaborated architectural structure of clothing (Kim 2012a, b). Dries van Noten, another world famous designer, presented *Hanbok*-inspired designs in his Fall-Winter 2012 Collection (Korea.net 2014). One of the design details used in his collection was *Dong-Jung* (the white paper neck band) combined with printing effects (Korea.net 2014). Korean *Hanbok* designer Kim Hye-Soon was the collaborator in this collection (Kim 2012a, b). From both Caroline Herrera and Dries van Noten's *Hanbok* inspired collections, it is evident that Korean aboriginal fashion has strong potential to be used as a unique inspiration source of materials and techniques as well as styles for the global fashion industry. It also shows that global runway presentation of Korean aboriginal inspired collections helps Korean fashion designers to create new opportunities to collaborate with global designers, giving them the chance to be exposed to global audiences.

3.2.1 Branded Fashion Companies

LIE SANGBONG

Designer Lie Sang Bong founded his fashion brand LIE SANGBONG in Seoul, Korea in 1985 (LIE SANGBONG 2015). He has been featuring his collections in multiple fashion runways in Paris starting in 2002 and presenting in New York Fashion Week from the Fall-Winter 2014 shows (Mercedes-Benz Fashion Week n.d.). He was the first designer to introduce *Hangul* calligraphy as aboriginal design motifs through his Seoul exhibition titled "*Hangul* walks on the moonlight" to global audiences (Mercedes-Benz Fashion Week n.d.). Lie Sang

Fig. 16 Designer Lie Sang Bong and his collection presented in Paris Fashion Week, Spring-Summer 2013 prêt-a-porte collection (LIE SANGBONG 2015)

Bong constantly reflects Korean aboriginal inspirations in his collections to promote them in the global market. For instance, he uses *Mosi* (ramie) and natural herbal dyeing methods (e.g., gardenia seeds, omija) (LIE SANGBONG 2015). The themes of his two recent collections were *"Bojagi,"* the Korean traditional patch work in Spring-Summer 2016 and *"oriental ink"* in the Fall-Winter 2015 Collection (LIE SANGBONG 2015). For both of these collections, traditional Korean patterns and motifs were transformed to contemporary styles and designs using innovative design technologies such as digital printing and laser cutting (Mercedes-Benz Fashion Week n.d.). Designer Lie makes his efforts to maximize the usage of global runways to promote Korean aboriginal fashion in the global fashion industry; see Fig. 16.

Troa

Designer Han Song made his brand Troa special with the exclusive use of *Hanji* (paper mulberry) blended fabrics and Korean traditional dyeing methods (e.g., Korean ink, indigo, soupberry tree) (Troa 2015). Troa's collection caters not only to luxury women's ready-to-wear, but also to casual denim lines, so the brand makes a distinctive contribution to expanding the usage of Korean traditional fashion components for larger target audiences (Troa 2015). Since Troa's debut in the Spring-Summer 2012 Seoul Fashion Week, the brand has played a bigger role in the promotion of Korean aboriginal materials and technique ins the global market (Kim 2011). In the debut collection, natural dyed *Hanji* denim pants and T-shirt collections were featured, which were fitted into the theme of the show "Be Kind to Earth" (Kim 2011). Troa is currently being sold to top global fashion retailers such as Barneys New York, Harvey Nichols, and Ten Corso Como

Fig. 17 Designer Han Song of Troa and Troa's *Hanji* collection presented in 2012 SS Seoul Fashion Week (Troa 2015)

(Troa 2015). Hence, the brand contributes to creating awareness of the sustainability of Korean aboriginal fashion in the global market; see Fig. 17.

Isae

Designer Chung Kyung-Ah of fashion brand Isae is one of the few Korean designers who is committed to traditional Korean fabric and natural dyeing method development (Lee 2014a, b, c). Her strong business ethic in sustainability and is reflected in Isae's products and conceptual flagship stores (Lee 2014a, b, c). From Isae's debut in the Spring-Summer 2011 Seoul Fashion Week, the brand has been continuously presented in global runway shows and exhibitions such as Milano Fashion Week and Preview in China (Isae 2015). Through these global presentations, Isae promotes the sustainability and aesthetics of traditional Korean natural fabrics and dyeing methods to global audiences.

3.2.2 Art Institutions

There are two distinctive art galleries that contribute to promoting global awareness of Korean aboriginal fashion: BLANK SPACE and Natural Dying Culture Center. BLANK SPACE is a creative art space, located within LIE SANGBONG's flagship store in New York (LIE SANGBONG 2015). LIE SANGBONG collections are presented in the gallery's unique artistic platform in conjunction with fashion, together with regular art exhibitions every 6–8 weeks (LIE SANGBONG 2015). BLANK SPACE is a unique example of incorporating art and fashion to feature Korean aboriginal fashion in the global market; see Fig. 18.

Fig. 18 BLANK SPACE art gallery and LIE SANGBONG flagship store in New York (LIE SANGBONG 2015)

In 2005, the Natural Dyeing Culture Center was founded to promote traditional Korean natural dyeing culture and techniques (Natural Dyeing Culture Center 2015). It is composed of a museum, museum shop, and research institution, located in Naju city of the *Jeonnam* province (Korea Natural Dyeing Culture Center 2015). The Natural Dyeing Culture Center holds multiple training classes, exhibitions, and fashion shows serving approximately 60,000 global visitors, and over 15,000 people take training classes annually (Natural Dyeing Culture Center 2015). Since 2006, the center also hosts a natural dyeing textile design competition every year and oversees the technical advisor license exams of traditional Korean natural dyeing (Natural Dyeing Culture Center 2015). The Natural Dyeing

Fig. 19 Natural Dyeing Culture Center in Najoo city of *Jeonnam* province, Korea (Natural Dyeing Culture Center 2015)

Culture Center performs cultural ambassador's roles to introduce traditional Korean natural dyeing to global audiences (Natural Dyeing Culture Center 2015); see Fig. 19.

3.2.3 The Artisan

Mr. Jung Kwan Chae is an artisan of *Yeomsaekjang*, the 115th Important Intangible Cultural Asset of Korea, who specializes in traditional Korean natural dyeing methods (Cultural Heritage Administration n.d.d). He is one of the most active performing artisans who present multiple seminars and exhibitions in global locations (Natural Dyeing Culture Center 2015). For instance, he showcased his works in Japan in 2011 and Taiwan in 2014 (Jung 2015). To create global awareness in the general public, including foreign tourists, he hosts open year-round training sessions and exhibitions in his natural dyeing transmitting center (Jung 2015). Artisan Jung's works are currently published in one of the national high school art textbooks (Jung 2015); see Fig. 20.

Fig. 20 Jung, Kwan Chae, an artisan of *Yeomsaekjang*, the 115th Important Intangible Cultural Asset of Korea and his natural indigo dyeing works (Jung 2015)

4 Limitations and Solutions of Korean Aboriginal Fashion

There are limitations for Korean fashion brands and manufacturers to increasing their involvement in natural fabrics and dyeing method-related business, mainly because of the collapse of the traditional Korean dress market. More specifically, the collapse was caused by: (1) a lack of global awareness; (2) limited consumer demands, because of preference for lower prices and fads; (3) limited manufacturer capacities, because of heavy dependence on hand-craft production. As a result, there are currently just a few manufacturers capable of producing large volume orders requiring elaborated traditional Korean natural fabric making and natural dyeing techniques (Isae 2015). As a consequence, global luxury fashion brands experience difficulty growing their adoption of Korean aboriginal fashion, because of limited sourcing availabilities. To overcome these limitations, global awareness and demands for Korean aboriginal fashion must be created through global promotional opportunities such as global runway shows and exhibitions. Contemporary manufacturing methods and adapted traditional techniques should be developed to reduce hand-craft manufacturing processes. For instance, contemporary *Hanji* fabric production processes have enhanced productivities and qualities of products by replacing old hand looms with modern instruments adapted from traditional manufacturing techniques (Troa 2015).

Incorporation of technologies and traditional designs can be an innovative solution in differentiating Korean aboriginal fashion products. Some Korean art craftsman studios are constantly investing efforts to maintain competencies to produce high quality traditional textile products using innovative technologies. For instance, Whasoomok specializes in R&D of traditional Korean natural dyeing methods. The craftsman studio developed high quality natural dyeing techniques using low temperature vacuum extraction methods and natural digital printing ink production techniques (Whasoomok n.d.).

5 Future of Korean Aboriginal Fashion in the Global Fashion Industry

Despite the challenges mentioned above, considering the significance of constant Asian market growth, there is still an increasing number of global luxury brands reflecting Asian aboriginal inspiration in their pre-season (cruise) collections (Fernandez 2015; Ruth Styles 2015). Additionally, the eco-friendly product market size is growing 2.7 % annually (Oh et al. 2010). These circumstances are favorable for creating new opportunities for the fashion brands and manufacturers involved in businesses related to Korean aboriginal fashion. However, discontinuation of traditional Korean dressmaking cultures and techniques is still a threat to the future of Korean aboriginal fashion. Hence, designers and artisans who are dedicated to protecting Korean cultural heritage must be supported by organizations such as global luxury brands and governmental offices.

6 Conclusion

Korean aboriginal fashion presents a sustainable option in the use of natural materials and processes and design techniques. Fashion products can be innovated with aboriginal pop-culture and eco-friendly components, and these products facilitate sustainable fashion consumption for global consumers. Korean fashion designers, artisans, and professionals in art institutions have been dedicated to promoting Korean aboriginal arts and designs and creating global awareness of their sustainability and aesthetics. Global runways and exhibitions are an efficient and impactful method for creating global awareness and new business opportunities for Korean aboriginal fashion. Therefore, the fashion companies, which include global luxury fashion brands, Korean fashion brands, and manufacturers, must be aware of the sustainability and advantages of Korean aboriginal fashion. They also need to invest their efforts in increasing the adoption of traditional Korean inspirations in the global runway collections and strengthening manufacturers' capacities in large volume orders. Future study should be conducted towards determining the sustainability and benefits of other Asian countries' aboriginal fashions to foresee their future opportunities in the global market.

Appendix

Category	Name	Contact	Phone	Location	Website
Branded fashion company	Isae FNC	CEO/Chung, Kyung-A	82-2-763-6818	Seoul, Korea	www.isae.co.kr
	Lee Young Hee	CEO/Lee, Young-Hee	82-2-544-0630	Seoul, Korea	www.leeyounghee. co.kr/
	LIE SANGBONG	CEO/Lie Sang-Bong	82-2-553-3380	Seoul, Korea	www.liesangbong .com
	Troa	CEO/Han Song	201-961-2211	New York, USA	www.troaco.com
Art institution	BLANK SPACE	Director/Nana Lee	212-924-2025	New York, USA	www.blankspacear t.com/
	Natural Dyeing Culture Center	CEO/Kang, In-Kyu	82-61-335-0091	Iksan, Korea	www.naturaldyeing. or.kr/
Artisan	Jung, Kwan-Chae	Jung, Kwan-Chae	82-61-332-5359	Naju, Korea	www.Jungindigo. com

Category	Name	Contact	Phone	Location	Website
Art craftsman studio	Whasoomok	CEO/Kim, Hoo-Ja	82-54-337-7715	Seoul, Korea	www.Whasoomok .com

References

Ahn S-H (2014) Isae's Hansan-Mosi goes to Milano. http://www.fashionbiz.co.kr/TN/?cate=2&r ecom=2&idx=138445. Accessed 12 Oct 2015

Anderson K (2015) 10 things to know about Chanel's latest resort show. http://www. vogue.com/13263892/10-things-to-know-about-chanel-resort-2016-show. Accessed 17 Oct 2015

Bae Y-D (2003) Tradition and current meaning of hemp cloth as native product in Andong. http:// academic.naver.com/view.nhn?doc_id=11282288&dir_id=0&page=0&query=%EC%95 %88%EB%8F%99%ED%8F%AC&ndsCategoryId=10128&library=151. Accessed 12 Oct 2015

Childs P, Williams R-J (1997) An introduction to post-colonial theory. Prentice Hall, London, p 211

Choi J-I, Chung Y-J, Kang D-I, Lee K-S, Lee J-W (2012) Effect of radiation on disinfection and mechanical properties of Korean traditional paper Hanji. Radiat Phys Chem 81(8):1051–1054

Craik J (1994) The face of fashion. Cult Stud Fashion 199:10

Cultural Heritage Administration (n.d.a) Weaving Chungyang Chun-Po. http://www.cha. go.kr/korea/heritage/search/Culresult_Db_View.jsp?mc=NS_04_03_01&VdkVgw Key=22,00250000,34. Accessed 17 Oct 2015

Cultural Heritage Administration (n.d.b) Weaving Hansan-Mosi. http://www.cha.go.kr/korea/her-itage/search/Culresult_Db_View.jsp?mc=NS_04_03_01&VdkVgwKey=17,00140000,34&fl ag=Y. Accessed 22 Oct 2015

Cultural Heritage Administration (n.d.c) Andong-Po. http://www.cha.go.kr/korea/heritage/ search/Culresult_Db_View.jsp?mc=NS_04_03_01&VdkVgwKey=22,00010000,37. Accessed 17 Oct 2015

Cultural Heritage Administration (n.d.d) Natural dyeing expert. http://www.cha.go.kr/korea/herit-age/search/Culresult_Db_View.jsp?mc=NS_04_03_01&VdkVgwKey=17,01150000,36&fla g=Y. Accessed 5 Oct 2015

Cultural Heritage Administration (n.d.e). Nu-Bi Jang. http://www.cha.go.kr/korea/heritage/ search/Culresult_Db_View.jsp?mc=NS_04_03_01&VdkVgwKey=17,01070000,37. Accessed 17 Oct 2015

Cultural Heritage Administration (n.d.f) Weaving Moomyung. http://www.cha.go.kr/korea/herit-age/search/Culresult_Db_View.jsp?mc=NS_04_03_01&VdkVgwKey=17,00280000,36&fla g=Y. Accessed 21 Oct 2015

Cultural Heritage Administration (n.d.g) Weaving Myung-Joo. http://www.cha.go.kr/korea/herit-age/search/Culresult_Db_View.jsp?mc=NS_04_03_01&VdkVgwKey=17,00870000,37&fla g=Y. Accessed 22 Oct 2015

Doopedia (n.d.a) Andongpo. http://terms.naver.com/entry.nhn?docId=1178953&cid=40942&cat egoryId=32092. Accessed 17 Oct 2015

Doopedia (n.d.b) Chun Po. http://terms.naver.com/entry.nhn?docId=1197460&cid=40942&cate goryID=33065. Accessed 22 Oct 2015

Doopedia (n.d.c) Myung-Ju. http://terms.naver.com/entry.nhn?docId=1093270&cid=40942&cat egoryId=31891. Accessed 13 Oct 2015

E-daily News (2013) Ssang-Young's CEO, Kang-Hoon Kim. http://www.edaily.co.kr/news/New sRead.edy?SCD=JC61&newsid=01128326602875832&DCD=A00306&OutLnkChk=Y. Accessed 7 Aug 2015, 17 Oct 2015

Eller J (1997) Ethnicity culture and the past. Mich Q Rev, vol 36

Evans C (1997) Street style, subculture and subversion. Costume 31:79

Fernandez C (2015) Gucci to show resort collection in New York. http://fashionista.com/2015/03/gucci-cruise-collection-new-york. Accessed 22 Oct 2015

Flugel J-C (1969) The Psychology of clothes. New York: International Universities Press, Inc., 129–130

Geczy A (2013) Fashion and orientalism: dress, textiles and culture from the 17th to the 21st century. Bloomsbury, London, pp 183–186

Germ K-S, Delong M-R, Revell M et al (1992) Korean traditional dress as an expression of heritage. Dress 19(1):57–68

Grayson J-H (1992) The accommodation of Korean folk religion to the religious forms of Buddhism: an example of reverse syncretism. Asian Folklore Stud 51(2):199–217

H&M (2013) H&M 2012 annual report. http://about.hm.com/content/dam/hm/about/documents/en/Annual%20Report/Annual-Report-2012_en.pdf. Accessed 17 Oct 2015

Han S-H (2013) Inheritance of traditional Myung-Joo. http://dlps.nanet.go.kr/DlibViewer.do?cn=KDMT1201415870&sysid=nhn. Accessed 18 Oct 2015

Higgins E-M, Eicher B-J (1995) Dress and identity. Fairchild Publication, New York

Isae (2015) ISAE Company Report. Seoul, Korea: Lee, M.

Isae Blog (2011) Isae's Chun-Po jacket http://blog.naver.com/itbit/40131625684. Accessed 17Oct 2015

Jang M-J (2011) Designer Lie Sang Bong introduces Korean traditional fashion through his Hanji lingerie collection. http://m.breaknews.com/a.html?uid=191556. Accessed 17 Oct 2015

Jeon Y-H (2011) A study on traditional beauty of Korean Bojagi—focused on the Bojagi and patchwork wrapping cloth of the Joseon dynasty. http://academic.naver.com/view.nhn?doc_id=45956427&dir_id=1&field=0&unFold=false&gk_adt=0&sort=0&qvt=1&query=%EB%3%B4%EC%9E%90%EA%B8%B0&gk_qvt=0&citedSearch=false&page.page=1&ndsCategoryId=10203&library=82. Accessed 11 Oct 2015

Jeong M-J, Kang K-Y, Bacher M-K, Kim H-J, Jo B-M, Potthast A (2014) Deterioration of ancient cellulose paper, Hanji: evaluation of paper permanence. Cellulose 21(6):4621–4632

Joy A (2015) Fast fashion, luxury brands, and sustainability http://www.europeanfinancialreview.com/?p=4589. Accessed Aug 7 2015

Joy A, Sherry J-F, Venkatesh A, Wang J, Chan R (2012) Fast fashion, sustainability, and the ethical appeal of luxury brands. Fashion Theory 16(3):273–296

Ju S-H (2011) Study on formative effect of the nubi. http://academic.naver.com/view.nhn?doc_id=75653203&dir_id=1&field=0&unFold=false&gk_adt=0&sort=0&qvt=1&query=%EB%88%84%EB%B9%84&gk_qvt=0&citedSearch=false&page.page=1&ndsCategoryId=10313&library=61. Accessed 17 Oct 2015

Jung K-C (2015) Traditional Korean natural dyeing methods. Najoo, Korea: Jung K.

Kim B-Y (2006a) Obangsaek, an expression of the Korean spirit. http://www.knutimes.com/news/article.html?no=526. Accessed 19 Oct 2015

Kim D-K (2006b) The natural environment control system of Korean traditional architecture: comparison with Korean contemporary architecture. Build Environ 41(12):1905–1912

Kim H-R (2011) Denim gets the Hanji treatment. http://www.koreaherald.com/view.php?ud=20110426000689. Accessed 17 Oct 2015

Kim H-K (2012a) K-fashion: wearing a new future. Korean Culture 7:13–94

Kim B-Y (2012b) Analyze work processes of Myungjoo, Mosi, and Chunpo. http://www.riss.kr/search/download/FullTextDownload.do?control_no=b9ad68775531a669ffe0bdc3ef48d419&p_mat_type=be54d9b8bc7cdb09&p_submat_type=f1a8c7a1de0e08b8&fulltext_kind=a8cb3aaead67ab5b&t_gubun=&convertFlag=&naverYN=&outLink=N-refer-&colName=bib_t&DDODFlag=&loginFlag=1&url_type=&query=. Accessed 17 Oct 2015

Kim J-H (2015) Korea needs to form Asian fashion league. http://www.koreatimes.co.kr/www/news/culture/2015/11/199_189977.html. Accessed 22 Oct 2015

Korea.net (2012) Five Korean traditional colors shine in New York http://www.korea.net/NewsFocus/Culture/view?articleId=102395. Accessed 22 Oct 2015

Korea.net (2014) Hanbok embodies age-old philosophies http://www.korea.net/NewsFocus/Culture/view?articleId=117325. Accessed 15 Oct 2015

Korea.net (n.d.a) Hansan ramie fabric cultural festival 2012. http://www.korea.net/Events/Festivals/view?articleId=4489. Accessed 17 Oct 2015

Lee J-E (2012) Fabric applied figurative characteristics of traditional nubi. http://academic.naver.com/view.nhn?doc_id=56547192&dir_id=1&field=0&unFold=false&gk_adt=0&sort=0&qvt=1&query=%EB%88%84%EB%B9%84&gk_qvt=0&citedSearch=false&page.page=1&ndsCategoryId=10204&library=26. Accessed 17 Oct 2015

Lee C-Y (2014a) Isae. http://www.apparelnews.co.kr/print.php?tbl=paper_news&uid=72753&sbj=%C0%CC%BD%B4%BA%EA%B7%A3%B5%E5%20-%. Accessed 11 Oct 2015

Lee S-M (2014b) Creative industry helps Hangul spread. http://koreajoongangdaily.joins.com/news/article/Article.aspx?aid=2989710. Accessed 18 Nov 2015

Lee Y-E (2014c) Designer Moon, Kwang Ja's Moomyung 2. http://www.wowtv.co.kr/newscenter/news/view.asp?bcode=T30001000&artid=A201408270491. Accessed 11 Oct 2015

Lee Young Hee (2015) Lee Young Hee press kit. Seoul, Korea: Lee A

Lie Sang Bong (2015) Lie Sang Bong company report. Seoul, Korea: Lee N

LIE SANGBONG (2015) LIE SANGBONG Brand Press Kit. New York, NY: Lee N

Lipovetsky G (1994) The Empire of fashion. Dressing modern democracy. Princeton University Press, Princeton, p 3

Maynard M (2000) Grassroots style: re-evaluating Australian fashion and aboriginal art in the 1970s and 1980s. J Des Hist 13(2):137–150

Mercedes-Benz Fashion Week (n.d.) Lie Sangbong. http://mbfashionweek.com/designers/lie-Sangbong. Accessed 19 Oct 2015

Natural Dyeing Culture Center (2007a) Natural dyeing. http://www.naturaldyeing.or.kr/xe/dyeing01/414. Accessed 5 Nov 2015

Natural Dyeing Culture Center (2007b) Types of natural dyeing materials. http://www.naturaldyeing.or.kr/xe/dyeing04/444. Accessed 5 Nov 2015

Natural Dyeing Culture Center (2007c) Persimmon dyeing. http://www.naturaldyeing.or.kr/xe/dyeing04/438. Accessed 21 Oct 2015

Natural Dyeing Culture Center (2007d) Safflower dyeing. http://www.naturaldyeing.or.kr/xe/dyeing04/441. Accessed 15 Oct 2015

Natural Dyeing Culture Center (2015) Natural dyeing culture center. Najoo, Korea: Kim D.

Oh D-H, Na H-J, Jeon Y-H, Jeong W-Y, An S-K (2010) A study on the physical properties of woven fabric with Hanji yarn. Text Sci Eng Korea 47(4):272–278

Olsen K (2011) Marni's mix: Consuelo Castiglioni makes jewelry out of records. http://www.vogue.com/869380/marnis-mix-consuelo-castiglioni-makes-jewelry-out-of-records/. Accessed 17 Oct 2015

Park H-Y (2015) Han-Bok in Boang-Saek. http://news.joins.com/article/18868554. Accessed 17 Oct 2015

Phelps N (2010) Spring 2011 ready-to-wear. Caroline Herrera. http://www.vogue.com/fashion-shows/spring-2011-ready-to-wear/carolina-herrera. Accessed 17 Oct 2015

Plus Korea Times (2008) History of Korean ink. http://www.pluskorea.net/sub_read.html?uid=11226. Accessed 25 Oct 2015

Roach M-E, & Musa K (1980) New perspectives on the history of western dress. New York: Nutri Guides,Inc

Roberts C, Huh D-H (1998) Rapt in colour. Powerhouse Publishing, Sydney

Ruth Styles (2015) The Korea-fication of high fashion: as Chanel launches its new cruise collection in Seoul, why designers are all looking east. http://www.dailymail.co.uk/femail/article-3071745/The-Korea-fication-high-fashion-Chanel-launches-new-cruise-collection-Seoul-designers-looking-east.html. Accessed 22 Oct 2015

Schapiro M (1980) Style: form and content. Aesthetics Today, pp 138–141

See A (1986) Apparduri, the social life of things. Commodities in cultural perspectives. Cambridge University Press, Cambridge

Shim Y-O, Park W-M (2003) Naju Saegolnayee. National Research Institute of Cultural Heritage, Seoul, Korea

Shin M-J, Cassidy T, Moore E-M (2011) Cultural reinvention for traditional Korean Bojagi. Int J Fashion Des Technol Educ 4(3):213–223

Suk C-S (1974) Traditional dress in survey of Korean arts. National Academy of Arts, Seoul, p 446

The economist (2013) How was Hangul invented? http://www.economist.com/blogs/economist-explains/2013/10/economist-explains-7. Accessed 18 Nov 2015

Trao (2015) Troa Press Kit. New York: Han, S. The Academy of Korean Studies. (n.d.) Moo-Myung. http://terms.naver.com/entry.nhn?docId=554262&cid=46671&categoryId=46671. Accessed 17 Oct 2015

UNESCO (n.d.) Memory of the world. http://www.unesco.org/new/en/communication-and-information/flagship-project-activities/memory-of-the-world/register/access-by-region-and-country/asia-and-the-pacific/republic-of-korea. Accessed 15 July 2015

Wang C (2012) Every single piece from the Marni for H&M collection. Recycled plastic bottle and wooden heeled shoes. http://www.refinery29.com/marni-h-and-m-collection#slide. Accessed 17 Oct 2015

Whasoomok (2015) Whasoomok's employee training manual. Seoul, Korea: Kim H

Whasoomok (n.d.) Introduction of company. http://www.whasoomok.com/ Accessed 13 Oct 2015

Xu Y, Wong M, Yang J, Ye Z, Jiang P, Zheng S (2011) Dynamics of carbon accumulation during the Fast growth period of bamboo plant. Bot Rev 77(3):287–295

Yoon J-Y, Yim E-H (2015) Symbolic meanings of women's dress on Korean film, Madame Freedom from the fifties. Fashion Text 2(1):1–15

Sustainability and Cultural Identity of the Fashion Product

Marlena Pop

Abstract Sustainability as a current general requirement consists in applying the principle by which all societal systems must be less dependent on resource use and products must last longer, thus imposing a slowdown in production and consumption and an increase in value addition and product customization. This step is essential because creative industries are constructed on the basis of the concept of surplus value of the product which may be a material constitutive, based on superior technology and cultural identity. In the creative interior of this artistic and technological area, the cultural identity of the product is defined by the cultural archetype or pattern which transcends the primary message and reveals, through the symbolic qualities of decorative elements and of ancient artistic techniques a whole specific universe. Studying cultural archetypes and patterns which define Romanian spirituality, expressed through symbolic elements and motifs of garb with ritual role, highlights elements of symbolic anthropology expressed through avant-garde concepts of modern product design. The transition from culture theory to artistic practice is made by deciphering the imaginary, the functions of archetypes and by understanding myths. Thus, the chapter argues the thesis that cultural identity sells any product provided the authentic intrinsic cultural value is respected, defined and promoted, because the European cultural economy is not only necessary and a top strategy—it is also a dynamic multicultural reality directed towards sustainability of heritage values.

Keywords Creative industry · Cultural identity · Fashion · Heritage value · Sustainability

M. Pop (✉)
National Research and Development Institute of Textile and Leather,
Bucharest, Romania
e-mail: pop_marlen@yahoo.ca

© Springer Science+Business Media Singapore 2016
M.A. Gardetti and S.S. Muthu (eds.), *Ethnic Fashion*, Environmental Footprints and Eco-design of Products and Processes, DOI 10.1007/978-981-10-0765-1_4

83

1 Introduction

The designers' approach to sustainable development implies socio-economic and individual changes in the business environment, with the purpose of providing measurable benefits in terms of global sustainability. From a humanistic perspective, the product of creative industries has a cultural identity, which means creativity, innovation and participation in a conceptual style and theme. In the fashion industry, this creative process occurs in man's natural rhythm and is found in what is called slow fashion, the opposite of *fast fashion*, where all processes, from product design to sale are very fast, occurring in the rhythm imposed by the computer. "Slow fashion is a better design, production, consumption and life, by combining ideas of renewal cycles and nature's evolution with those related to values and traditions" (Flatcher 2008, p. 68). Query ID="Q2" Text="Kindly note that there are 17 'footnotes' which in fact point to references. The footnote numbers in the text and the footnotes themselves should be removed and replaced by reference indicators in the text referring to items in the list of references. Please check and confirm."

Embedding the concepts of clean economy and, since 1987, sustainability, but always oriented towards creativity, the fashion industry has introduced the European cultural paradigm in research on product and brand trends, demonstrating its innovative quality since the 1970s, also adapting it to the requirements of sustainability in the last 20 years. The fashion trends of the last 30 years, as well as shows of the large luxury brands, particularly French and American, feature motifs, symbols, and palettes of cultural and artistic or artisanal traditions of the world's most expressive peoples. Thus, the participation of art, psychology, history, or anthropology, and therefore the humanities is increasingly more customized to societal interests of fashion as a major social phenomenon, regarding both creating the fashion product and developing the creative industries in the twenty-first century.

In a possible taxonomy of textile arts (Fig. 1) it can be observed where fashion, costume art, and fashion design lie in the panoply of contemporary fine and decorative arts. It can be seen that much textile art contributes to the cultural development of fashion, including the art of costume (in all kinds of manifestations, from restoration to costume performing), art fabrics, textile prints, embroidery, patchwork, collage, and mixed 2D textile design. This contribution is direct, practical, and applicative, through artistic creation. However, in turn, any artistic production is increasingly richer if it resorts to the humanistic foundation of artistic discourse involving the elements of art history, anthropology, psychology, semiotics, philosophy, sociology, etc. In particular, fashion is determined in its evolution by studies, research, and innovative concepts of a school of thought at a given time. Any fashion product, clothing or object, such as clothing accessories and the whole range of life-style objects, even including houses, cars and the urban environment, becomes a sustainable cultural product only to the extent to which it is developed on the education—research axis in every industrial field, with proper research, both technical and humanistic, oriented towards sustainable development.

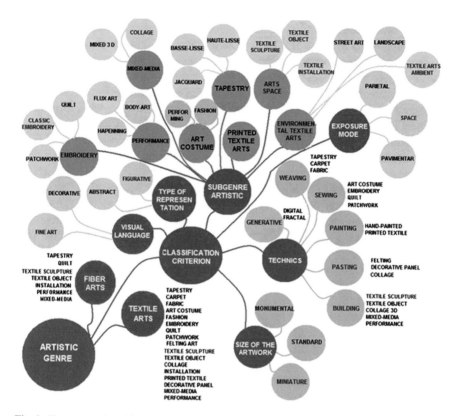

Fig. 1 Taxonomy of textile arts—author sketch

Differences between two types of thinking in design are highlighted: classic design and sustainable cultural design, present only in the educational-empathic and thematic level of thinking in design, in the stage of research and outlining the idea of a product, which demonstrates that the principle of cultural sustainability and the practice of cultural design in fashion, are an essential condition for the originality of the product.

In this context, the following assertion can be put forth, comprising both thesis and antithesis: *it is the cultural sustainability of the product that provides its originality or the originality arising from the identity of a culture is the one that provides cultural sustainability of a fashion or life style product and of cultural industries in general.* The synthesis of these statements proves that *cultural sustainability of a product is defined by the originality arising from the identity of a culture that becomes a theme of creativity through its endurance over time.*

In a globalized economy there is an increasing need for unique-uniqueness, new-innovative identity, artistic or scientific, and this manifests in the production of goods through increasingly imaginative ideas, solutions, and concepts. The solutions offered globally are not "globalist" clichés, but claim to have cultural identity: brand identity, author identity, a recognizable cultural identity, ethnic, historical, aboriginal identity, etc. Thus, from an imagistic point of view, the product of creative industries has an increasingly evident cultural identity. Artistic identity, an identity of postmodern artistic fashion, in a postindustrial era, is the spearhead of this identity trend. However, there is a current of ethnic identity of the term in the sense of aboriginal. This means access to a treasure of civilization as a source for new creative ideas, innovation, and belonging to an archetypal ideational style and theme, mythical but historically particular, as artists like to say: a local identity with universal value. In this context, cultural sustainability of Romanian creative industries consists of developing a cultural construct in which the sustainability of cultural landmarks of creativity, defined by complex structures of the Romanian individual and collective imaginary, would be more and more visible. This "construct" can be a cultural work instrument used by artists, designers, and craftsmen.

Practice in the visual arts is a form of understanding human cognitive processes, exploring diverse backgrounds, and as much of the collective consciousness of the world as possible. The results of artistic research and practice are so diverse that today, in the world of information, they can be found anywhere, from utilitarian objects to visual text, image, blogging, and hacking.

With this enormous diversity of opportunities, beyond visual and cultural clichés belonging to a globalized world there emerges the expression of the creative self, of otherness, of the magic or the traditional, of an imaginary that tries to express typologies of identity, other than the already existing ones of the moment.

One of the easiest connections with the self is identification by belonging to a national history and culture of the creative self, in that the mythical universe is personal and therefore unique, but included in a collective imaginary, a space of being between identity and otherness in the mythical world in which we develop as people first and then as artists or other sources of self expression. From the archetypal collective unconscious, poetic or visual images emerge, proving the specificity of the imaginary of a certain culture. In the twentieth century, the forefront research of Gaston Bachelard, Henry Corbin, Mircea Eliade and, above all, Gilbert Durand revealed the specific nature, mechanisms of generation, operation and transmission of the imagination, as well as the synaesthetic, symbolic and logical relations it maintains with perceptions, ideas, and memory, also highlighting the role of imagination and its functions within the individual psyche, the collective one, and that of social praxis. A particular subdomain, such as artistic imaginary research, aims to identify archetypes of fascination which continuously feed the universe of artistic images. In his study of the historical evolution of the imaginary, G. Durand develops an analysis of methodology, considered a model of mythology evolution in a given cultural period.

Demonstrating the functions and values of the imaginary, the author highlights both the creative dynamics of the imaginary and the symbolic poignancy to which the subject adheres, highlighting three important features: *playful aesthetic*, *cognitive*, and *practical-established* function. The definition given by the Alsatian philosopher Jean-Jacques Wunenburger in his review work *The Imaginary* to this whole complex ensemble of visual images and "linguistic ones—metaphor, symbol, story—forming coherent and dynamic structures and revealing a symbolic function, in terms of a combination of meanings of a literal and figurative nature" (Wunenburger 2003, p. 56), is operational in all artistic productions.

All this research on the imaginary finds its applicative demonstration in the originality and identity of the visual artwork, the artistic product, or the design concept. In turn, analysis of identity in art has drawn the interest of analysts such as Pierre Bourdieu through the "language of symbolic power" of Jonathan Friedman or, more recently, Thomas McEvilley. Thus, in his book *Cultural Identity and Global Process*, Friedman unravels the societal mechanisms of culture, saying that "The construction of identity in relation to cultural creativity involves local social and individual mechanisms that it is necessary to understand if we are to fully grasp the importance of the local/global articulation in the production of culture" (Friedman 1994, p. 27). He also says that, from an anthropological perspective, the levels of variation of the "cultural identity" term are distinctive. Thus we have cultural identity roots, a modern ethnicity, a traditional one, but also a contemporary one, defined by lifestyle. Analyzing the consumers of the twentieth century, Bourdieu can only define them in terms of their cultural identity, because only this can build cultural niches in the world of consumer goods. In the same spirit, McEvilley supports the idea that shows us that the expression of multiple cultural identities is in fact a necessary "global pluralisation." The American philosopher also offers a solution to the postmodern crisis of cultural identity in the visual arts: this may be "imaginative integration" (McEvilley 1992, p. 43).

Well before the American philosopher, at the beginning of the twentieth century, in *Myths, Dreams and Mysteries*, defining inter-cultural dialogue and a "new humanism" (Eliade 1962, p. 241), one of the greatest Romanian philosophers claims that modern man may escape the "terror of history" by learning from traditional cultures. Many traditional cultures have myths about the end of their world or civilization. However, these myths do not succeed "in paralysing either Life or Culture" (Eliade 1962, p. 243). Mircea Eliade said half a century ago that "everything that is happening today in European culture leads us to believe that soon we should see a restoration of symbolism as an instrument of knowledge" (Eliade 1992, p. 44). The development of visual semiotic of archetypology within ethnological and anthropological studies and the involvement of cultural studies in the analysis of global phenomena in certain socio-humanistic areas have enabled researchers to go from historicism to cultural studies of visual anthropology. Thus, the forecast of the Romanian philosopher came true: "Showing the cradle of a people is of much higher interest than deciphering a medieval manuscript. It is no longer a great "glory" to create history. To belong to a native "race" is much more precious" (Eliade 1992, p. 45).

In this context, cultural sustainability can be investigated through arts, myths, and archetypes belonging to meanings of magical thinking in whose crucible initial forms of pre-logic were decanted, in the sense assigned by Levy-Bruhl and Vilfredo Pareto, which cannot be subjected to binary limitations. Symbols and metonymy, representations and metaphors escape from noncontradiction laws, from molds and imitation, and allow the union of opposites (*conjunctio oppositorum*) in evolved forms of postmodern thinking, where cultural industries open an innovative refreshing perspective on the arts.

Why arts? Because openness to spirituality, to conscience happened when primitive man carved testimonies of the times he lived in on the walls of caves, even if he was not aware of his gesture. Why myths? Beyond the fact that all of us like stories and myth implies the story of Genesis, neurotheology findings (through the research of geneticist Dean Hamer 2004) claim the biological existence of a *gene of God* (VMAT2) on the canvas of which all our metaphysical experiences are found. Why archetypes? By virtue of the common denominator that animates the cognitive activity of humanity, the peak of individual experiences are put together in a collective unconscious, ensuring humanity's cultural landmarks and perenniality.

However, artistic imagination is revealed not only through explicit visual semiotics and rhetoric but also through ineffable emotional transgression, which is much more difficult to decode by psychologists, philosophers, and aestheticians. Cultural identity in the visual arts is defined by the archetype that transcends artistic work, beyond the artist's personal myths, beyond semiotic myths of cultural fashion; basically, by transgressing the direct message of the artistic work and the intentionality of the artist, the archetype reveals only the original reference point of our historical and spiritual being. Following the scientific line of CG Jung in defining the archetype, but also the philosophy of Mircea Eliade, who said that "symbol, myth, archetypal image are related to the substance of spiritual life, that we can hide, mutilate and degrade them, but never remove them" (Jung 2003, p. 143), we are trying to decode that particular Romanian mythical archetypal syntax which is expressed in our contemporary art as well. For the Romanian space, the concepts created by Lucian Blaga are important because he defines myth in close relation to the mystery, as are the concepts by Mircea Eliade for whom myth is the absolute sacredness. Mythology, as the science of myths, which demonstrates the ability of each country to represent its own imaginary symbolic representation is also, according to Romulus Vulcănescu, "an ethnic system of myths", and a system "reflecting a certain mythical view, a Pantheon and an oral mythical literature which is distinct by its essential characters" (Vulcanescu 1987, p. 92). The transition from culture theory to artistic practice is made by deciphering the imaginary, the functions of archetypes and by understanding myths. Which is the structure of the Romanian mythology? Fundamental myths: *Myth of spirituality—Miorita ("The Little Ewe"), Myth of ethnogenesis–Baba Dochia ("The Old Dokia"), Myth of the creation–Mesterul Manole ("Master Manole"), Erotic myth–Zburatorul ("Incubus"), Cosmogonic and astral myths*: Creation of Heaven, Tree of Heaven, Vamesii vazduhului, The Holy Sun, Cosmic Egg, Sun Wheel, Calusarii (Calucenii), Zburatorul (Incubus), Fat-Frumos (Prince Charming) and his counterpart Ileana Cosanzeana, the Sun and the Moon, The Fire of Life, Joimaritele (Old maids), Sanzienele (Fairies), Stele logostele (Stars), The Morning and the Evening Star, The Milky Way, etc.

Geogonic myths: Mother Earth, The Good Twin God and the Evil Twin God, The Other World, The Old Mother, White Island, Apa Sambetei (Saturday's Water), The White World, Uriesul etc.

Myths of heroes: Zamolxis, The Good Twin God, Mistricean, Decebalus, Trajan, Bessarabians, Stefan the Great, Michael the Brave, Avram Iancu, Master Manole, The Novac brothers, Pintea the Brave, Gruia, Iorgovan, and folk characters Pacala, Pepelea.

Myth is the expression of an encoded message whose decoding key we must find. Myths always include basic symbols, narratively exploiting the power of the symbol to make transparent the cryptic dimensions of reality.

There are two fundamental dimensions of the symbol: *gnoseologic* and *ontologic*. Both appear in Eliade's analysis without being explicitly denominated as such. His effort focuses on the disclosure of internal coherence of symbolic systems. However, once intelligibility is guaranteed, the matter of the *meaning* communicated by the symbol (related to myth) gives way to *efficiency* (power) communicated by the symbol (related to rite). Just by being universal (objective) structures of the human spirit, archetypes become cultural "symbols." "Because of their inherence in any important human enterprise, they gain aesthetic value" (Eliade 1978, p. 79).

Symbolic ornaments reflect customs, myths, personifications, beliefs, figurative representations of institutions, aspects of folk knowledge, and practices. Archetypal symbols found as dominant motifs in ancient textile crafts are divided in the following groups: *mythological, ethnographic-folkloric, religious, emblematic-heraldic* mythical symbols: "solar signs," "suns with little fir trees," "sun," "sun wheels," "sun and reel" ethnographic symbols: "fir tree leaf," "tree of life" whose significance is "eternal youth and life without death," "four-leaf clover" which stands for luck. religious symbols: "cross," "candlesticks" emblematic or heraldic symbols: "bison head"—Moldova, "vulture and crosses" Transylvania, "peace dove"—bride kerchief in Western and Southern Carpathians.

Romanian archetypal symbols are solar signs, tree of life, fir tree, red apple, four-leaf clover, cross, prosphora, bison head, vulture, and white doves. One of the most common motifs is the "zig-zag"—abstract, geomorphic, also known as *"calita-ocolita," "carnel"* or "waves," "calea ratacita," "coarnele berbecului," and "carligul ciobanului." Another type of "self-mirror," or mirror of macrosocial consciousness, is built by interpreting *cosmomorphic ornamentation* found in our "ethnocultural space" (Stanciulescu et al. 2008, p. 245).

The solar cult of our ancestors is found in solar archetypes expressed by the following motifs: solar wheels, representations of the heavenly bodies, and the succession of seasons. In addition to the sunlight cult, there are the cult of moonlight, star, and rainbow.

Romanian folk costume—priestly dignity. Analyzing archetypes and cultural patterns which define Romanian spirituality, the philosophical and artistic argument developed by analysis of Romanian myths and textile motifs also allow analysis of contemporary clothing patterns, as well as understanding and interpreting the traditional symbolism expressed through the Romanian avant garde concepts of textile design.

Celebration, ritual, or wedding clothes were *"bridges"* to eternal life based on the belief that clothes—especially shirts—are *"the house of man, of his body"* (Eliade 1978, p. 85).

A Romanian peasant blouse seems to us a manuscript with strange hiero-glyphs, a waist belt, through its motifs, an important ritual whose clothing purpose is derived, a head ornament, through the hierarchy of life of its wearers, seems nothing but an object of feminine coquetry (Image 1). Stylistic rigor is too severe to not precede aesthetics, decorativeness, and usefulness. It forces us to consider things from a different angle, from another perspective. Folk art chromatically reintroduces us into a world of white, basic color in Romanian art, especially in folk costume.

"A people in white. The choice of colour in ancient cultures always has a ritual significance" (Anghel 2003, p. 102). Color responds to a concept, it is assigned some value in the order of cosmic representation; its assimilation or rejection sep-arates communities in terms of ethnicity, just as distribution or right of jewelry hierarchically separate or place individuals within the family or clan. "Regardless of the motifs that separate ethnographic areas and which are, it seems, rudiments of tribal disintegration, in our country, white is unifying, a sign of ethnic particu-larities with underlying spirituality" (Anghel 1978, p. 175) (Image 2).

The *stylistic matrix* theory of Lucian Blaga is the one that gave new mean-ing to the concept of the cultural model in the mid-twentieth century. In *Horizon and Style* the Romanian philosopher says that the stylistic matrix "may be the permanent substrate for all the creations of a lifetime of an individual, at least in its essentials, similar almost to equivalence, to that of several individuals or of an entire nation or even part of mankind in the same era" (Blaga 1998, p. 42),

Image 1 Sacred symbols in Romanian mythical ornamentation. *Source* www.semnecusute.ro

Image 2 Queen Mary of Romania, Queen Mary and Princesses, Muscel peasant spinning, Princess Ileana, 1923 photo. *Source* www.lablouseroumaine.ro

referring to that immersion in the magic spirituality of cultural time and space of each person's childhood, which is unique, but subjected to the influence of collective, local, or global imaginary. Blaga's entire theoretical cultural construction developed in the style category, as an Archimedean point of support, where style is a "dominant and defining phenomenon of the human culture, an immanent space for any human creation" (Blaga 1998, p. 45), a theoretical construct that represents a valuable explanation of the societal cultural. These postulates lead to the understanding that the entire Romanian culture is basically a culture of memory, a material and spiritual memory. Cultural memory involves rationality of both cultural life and the unconscious archetype of this culture. Only by the unconscious archetypality of culture has the "stylistic matrix" emerged as a philosophical sign of Romanian cultural specificity.

The transfer of symbolic anthropology elements occurs in art and fashion. An archetypal landmark proves that any motif in traditional ornamental work is part of an entire regional mythology, which can be interpreted by merely referring to a repertoire of Romanian cultural motifs. We draw the conclusion that the motif in a textile work of art or fashion is a concrete spiritual archetype which manifests itself aesthetically and semantically as a valuable element of a mythos.

In the series of paintings called generically *La blouse romaine*, Henri Matisse transcribed these mythical signs (1935–1942). Elements of ancestral symbolic anthropology, characteristic of the Romanian folk costume, can be correctly transposed and reinterpreted if scientific methods of conceptual transfer are used at the level of visual language, or reinterpreted using high-level conceptual design methods.

Dacian spirituality has been known since ancient times, mainly dominated by a solar cult, generating an extraordinary inner strength (confirmed by the lack of fear of death). Dacian initiates had a profound state of communion with nature and even detachment from material things, which has been brought into focus nowadays by many artistic, literary, or philosophical representations. Of special note of Maramures culture in the artistic and philosophical discourse is the Merry Cemetery (Cimitirul Vesel) of Sapanta. It was built in the spirit of the Zamolxian religion of Dacians which imposed a state of joy at the funeral, sending off the deceased to a better world, but influenced by Slavonian chromatic of southern Slavonians who passed through Maramures.

In 2014, designer Lana Dumitru achieved more than a design concept, by creating a cultural fashion phenomenon by launching the "Merry Cemetery" collection followed by "Pink cemetery," inviting Romanian women to become vivid, colorful, and tonic as in the cemetery of Maramures (Image 3).

The cultural transfer of signs, motifs, and their symbolic, mythical, and ritual meanings is understood by means of the praxis of visual semiotics. Visual semiotics has the purpose of establishing or explaining the connections between a sign (signifier—the meaning, the signification of the clothing item) and the represented object (signified—the object and shape of clothing) in the context of a wide range of application areas. Transfer of cultural patterns and symbols through the fashion product takes place at different levels of stylization and visual reinterpretation. The experimental concepts of designer Andra Clitan prove the existence of four levels of stylization of Maramures motifs, from integration of a Maramures carpet motif in the tailoring of a modern clothing item, using a manual Maramures weaving style as raw material for a fashion piece (skirt with rose pattern), stylization of motif and its integration into the author's style (white blouse with Maramures lace motif) to the suggestion of the motif and, therefore, of the symbols it carries in a piece where the symbolic dominant is given by the structure of felt used in the winter "clop" (traditional Maramures hat). For the 2013 Berlin Fashion Festival, Andra Clitan presented some pieces in a different cultural concept, related to the rites of passage, wedding, and death. The pieces are presented in parallel and the stylistic processing resembles the figures and chromatics of Dacian tribes that later developed the traditional folk costume of Maramures. It is the straight figure and

Image 3 Lana Dumitru, *Pink Cemetery, Collection* s/s 2014. *Source* www.lanadumitru.ro

black and white chromatic, where white means death and black life, as in the early Dacian culture (Images 4 and 5).

At the opposite end we have rites of passage related to the wedding; a whole suite of female characters dominate this theme. For Romanians, the most important fashion piece is the Romanian traditional shirt (*IE*), *borangic*—the magic silk shirt (natural Romanian silk), with a rich ornamentation, which is an open book of beauty and popular wisdom that comes from the Neolithic proto-Dacian period. Many Romanian designers are fascinated by the richness of the Romanian shirt and many designer collections have taken and sometimes interpreted this piece of celebration costume, from YSL in 1972 to Oscar de la Renta in 2014. For the Moldavian Valentina Vidrascu, the Romanian shirt is poetry which brings romance

Image 4 Andra Clitan, "Solo Scriptura", Berlin Fashion Week 2013. *Source* www.ma-ra-mi.com

Image 5 Andra Clitan, "Solo Scriptura", Berlin Fashion Week 2013. *Source* www.ma-ra-mi.com

and spirituality. Her 2014 collection, *Ileana Cosanzeana,* takes us into the realm of Romanian tales, among fairies and fair maidens (Images 6 and 7).

The flight is another Dacian philosophical concept attesting to beliefs in immortality of this great European people with a monotheistic religion, their faith in Zamolxis, in times dominated by the polytheism of Greeks and Romans.

Image 6 Valentina Vidrascu, *Ileana Cosanzeana lux*, s/s 2012. *Source* www.valentinavidrascu.ro

Second year students of the National University of Arts, Fashion Department, tried
to render the spirituality and emotional state of the flight as a spiritual concept and
experience. The holistic informational method brings into question the collective
project, collaborative only in this experiment, but it can also be applied individu-
ally to increase the awareness of stages of integrating cultural information in an

Image 7 Valentina Vidrascu, *Stranger than Paradise*, f/w 2014–2015. *Source* www.valentinavi
drascu.ro

individual artistic project. A collective project, this time of an entire team, the second year students of Fashion, coordinated by Dr. Ioana Sanda Avram, was given a name that is a spiritual quintessence for the Romanians, "Flight" (Image 8). In this project, the entire creative approach of conceptual artwork in the spirit of the art of mainly costume and fashion can be deciphered, basically offering a unique

Image 8 "Zbor" (Flight) Collection, 2015, cordinator project UNA Bucharest, Dr. Ioana Sanda Avram. Image by permition of the coordinator

artistic state of mind, suggested by a Romanian archetypal symbol: Maiastra. We find this archetypal symbol both in the myth of "Zburatorul," the erotic myth of Romanians, in Romanian folklore, from the ballad of Master Manole, fairy tales such as "Eternal Youth and Life without Death" or "Praslea the Brave and the

Golden Apples", and in the arts: poetry—Eminescu, Blaga, Bacovia, etc., the famous sculpture of Constantin Brancusi, textile arts, etc. The archetype of flight accompanied the Romanian individual throughout his entire historical existence, as recounted in the drawings and motifs of Neolithic cultures, and also in ceramics, fabrics, encrusted wood, traditional clothing, artifacts that define the typology of traditional culture, as well as ancestral collective imaginary.

Interpretation of Romanian cultural patterns by a foreign designer, in the case of our example by the French designer Philippe Guillet, beyond the innovative show of the collection in the Romanian cultural environment at the end of 2011, is not an easy endeavor. The designer has interpreted the message of textile arts and folk costumes only at an aesthetic level, and that of techniques used for embroidery, lace, weaving or sheepskin coats only as manual work. Perhaps the French designer did not mean to send the symbolic and mythical message of Romanian spirituality, which does not forsake any of its components. When the reference of plastic semiosis is a folk costume, the ornamentation must not be separated from the tailoring style because the cultural identity of the product can be lost, and the stated purpose of such a collection, which is always "the valorization of cultural heritage," no longer exists, being destroyed by the cultural mixture. The affirmation is truly highlighted by making a bullfighter bolero jacket with bead embroidery and Bistrita motifs, English renaissance collars made of Maramures fabrics, or Spanish or French tailoring with some Romanian decorations. In these pieces, dominated by a multicultural message, Romanian cultural patterns become intransmissible and unidentifiable, losing themselves in the general expressiveness of the product, because the Spanish bolero is still a Spanish bolero, and its rich Bistrita ornamentation does nothing but aesthetically enrich this piece, saying nothing about its symbolism and origin. Through his collection "Preconceptions" the French designer has managed to prove only the existence of an active creative endeavor in Romanian crafts, ancient in its symbolic, mythical, and ritual structure, ready at any time to integrate as superior element in Romanian cultural products (Images 9 and 10).

Unlike this experiment of a foreign designer who wanted to highlight the richness of Romanian folk traditions, trying to mix the cultures of Europe so that the fashion public would more easily understand the message of Dacian culture, the experiment of a dissertation thesis of some Romanian students, in a Romanian university of art is much cleaner from the aesthetic standpoint and easier to read from the semiotic standpoint. Thus, in 2014, for their dissertation work, Emilia Tudoran and Iulia Ghenea analyzed the dialogue between the Moldavian folk celebration costume and the Renaissance collar from the medieval European costume. The result was the creation of an urban character, rich in plastic expressiveness and recognizable as Romanian identity (Image 11).

The motifs of Romanian folk costume are also found in the casual collections of Florin Dobre. The 2016s/s collection, dominated by the black/white drawings of Transylvanian motifs from the sub-Carpathian area of Fagaras, demonstrates that the Romanians wear with great pleasure contemporary clothing with elements or interpretations of Romanian motifs (Image 12).

Image 9 Philippe Guilet, "Prejudices" Collection, 2011, Embroidery of Bistrita and housings Straja, Suceava, Romania. *Source* www.100%.ro/fashion

Image 10   Philippe Guilet, "Prejudices" Collection, 2011, in the project 100 %ro and the master furrier Constantin Jurvale, Straja, Romania. *Source* www.100%.ro/fashion

Economic and cultural globalization, through its mechanisms of leveling life, consumption, and product recreation, has led, over the past 25 years, to the destruction of national identity in products and to an overestimation of international brands. The development of concepts such as sustainability in fashion, fast-fashion, slow-fashion, and ethic-fashion has recently brought back into question the need for national cultural identity, verbally and visually expressed through

Image 11 Emilia Tudoran and Iulia Ghenea, master UNA Bucharest, 2014, coordinator Ass. Prof. Dr. Paula Barbu. Image by permission of the coordinator

the product. Going back to the origins manifests in fashion concepts of the last two seasons through the interpretation of fashion themes related to the national heritage in very high cultural and artistic registers in terms of visual language. Conceptual design, through innovation and avant garde ideas, has the role of creating national cultural identity in fashion by re-evaluating cultural, spiritual, and

Image 12 Florin Dobre, Casual Collection, f/w 2015–2016. *Source* www.florindobre.ro

hand-made heritage, and the fashion industry must request these design products and services.

In conclusion, the analysis of the importance of culture as an element of originality of new products has become an increasingly studied issue, focusing, primarily, on promoting new societal values by fashion trends (primarily cultural values—arts, heritage, innovation) by bringing aboriginal cultural identity in through contemporary fashion product customization and continually promoting ethical fashion. Thus a whole arsenal of cultural sustainability is outlined, which needs a humanistic episteme to focus its scientific effort on finding new creative solutions.

The practical experience of the last few years in Romanian fashion design has been directed towards integrating mythical, cultural-artistic, identity elements into the product concept. It was noticed that the visual identity of a national culture is a positive influence on trendsetters, creating cultural emulation in the field, which is directly transmitted to the buyer and therefore to a market segment. Thus artistic endeavors of fashion designers who approach folklore, archetypes, or traditional cultural ideas become sustainable, as the strength of the world's cultural economy increases, which turns more and more from a goal and a top strategy into a present and future cultural reality.

Cultural sustainability of fashion proves to be a permanent invitation to an exercise of admiration: admiration for an entire traditional cultural heritage of the world that reveals at any given time the great richness of expression of humanity.

References

Anghel P (2003) Obarsie si perenitate. Editura Litera, Bucuresti
Blaga L (1998) Orizont şi stil, (Horizon & Style). Editura Humanitas, Bucharest
Eliade M (1962) Myth and reality. Harper&Row, Publishers.Inc, New York
Eliade M (1978) Myth and reality. Univers Press, Bucharest (in Romanian)
Eliade M (1992) Fragmentarium, Bucharest. Editura Humanitas, p 44
Flatcher K (2008) Sustainable fashion and textile. Design Journeys, UK, p 69
Friedman J (1994) Cultural identity and global process. Sage Publication Ltd., London
Jung CG (2003) Complete works 1. Archetypes and the Collective Unconscious. Trei Press, Bucharest, p 143 (in Romanian)
McEvilley T (1992) Art and otherness: crisis in cultural identity. Publisher Kingston, New York
Stanciulescu TD, Pop M, Grigoriu A (2008) Vestimentatia biofotonica - o sansa pentru sanatatea umana, Ed Performantica, Iaşi
Vulcanescu R (1987) Mitologie română. Editura Academiei, Bucuresti
Wunenburger JJ (2003) L'Imaginaire. Presses Universitaires de France, Collection "Que sais-je", Paris

Author Biography

Marlena Pop is a senior researcher in textile arts and fashion design of the *National Research and Development Institute for Textile and Leather* Bucharest, Romania. PhD in visual arts. She was for many years an associate professor at the *University of Arts and Design* in the Cluj-Napoca, Romania, and was textile and fashion publisher, for 5 years, of the *Textile Dialogue* magazine.

Badaga Ethnic (Aboriginal) Fashion as a Local Strength

H. Gurumallesh Prabu and G. Poorani

Abstract The Nilgiris in India has a fantastic history, tradition and culture. *Badagar* is a unique community inhabiting the Nilgiris and the people are commonly referred to as "Badagas". The badagas have their own distinctive and unique ethnic cultural practices. The customs governing their marriage, function, funeral and other events are exclusive to them. Poverty never existed among the badagas because of their sharing and caring social system. There was no need for the badagas to move out of the Nilgiris in the past, but in recent years a few badagas have moved from their villages (hattis) and migrated to other places in India and abroad to seek education, business and employment. The ethnography details about badagas were reported mainly by non-badagas in the past but, with recent technological progress, publication of research reports, books and documents is being arranged by native badagas. Badagas give importance to their past culture and are very keen to sustain it. As practiced by their ancestors, "Hethey" and "Hiriodaya", badagas are still preserving and practicing the traditional style of attire by wearing white dresses. The present chapter showcases the unique badaga system with special reference to the attire that has been followed over centuries as fashion. Irrespective of age, all badagas tend to wear traditional attire in the form of thundu, mundu, dhuppatti, seelai, mandarai, mandepattu, etc. on almost all social occasions and perform traditional badaga dances as a local strength and an ethnic fashion.

Keywords Attire · Badagas · Culture · Fashion

H. Gurumallesh Prabu (✉)
Department of Industrial Chemistry, Alagappa University, Karaikudi 630003, India
e-mail: hgprabhu@alagappauniversity.ac.in

G. Poorani
Department of Biotechnology, Kumaraguru College of Technology,
Coimbatore 641049, India
e-mail: gpoorani.biotech@gmail.com

© Springer Science+Business Media Singapore 2016
M.A. Gardetti and S.S. Muthu (eds.), *Ethnic Fashion*, Environmental Footprints
and Eco-design of Products and Processes, DOI 10.1007/978-981-10-0765-1_5

1 Introduction

The Nilgiris, one of the oldest mountain ranges, is located at 11°25′N 76°41′E, the tri-junction of Tamil Nadu, Kerala and Karnataka states in India. It has an area of 2565 km^2 with an elevation of 2789 m above mean sea level. The average temperature in summer is 6 °C (43 °F) and in winter is −12 °C (10 °F).[1] The Nilgiris is also known as the "Blue Mountains" because of the bluish smoky haze (Fig. 1) enveloping the hill areas (Fig. 2).

There are many reports available about the ethnic origin of the badagas. It is assumed that badagas inhabited the Nilgiris thousands of years ago, even before 8000 BC. It is reported that Buddhist monks entered the Nilgiris to spread among the badagas during the Mouriyan period. King Vishnuvarthana of Hoysala defeated the badaga King Kalaraja and the Nilgiris came under Hoysala in 1116 AD. Later, the Nilgiris was under the Vijayanagar Empire and, after the death of Tipu Sultan, it came under the British. A Portuguese priest by the name Ferreiri set foot in the jungles of the Nilgiris in 1602. The East India Company entered the Nilgiris around 1810. John Sullivan of England and Director of the East India Company went up the Nilgiris in 1819. He explored the plateau, introduced horticulture, created a lake and established summer residences at Ootacamund (now called Udhagamandalam), Ooty for short, which were considered as ideal for health, comfort and leisure. Ooty became the best choice for invalided officers and other Europeans seeking rest cures. Breeks in 1873 reported that the entire Nilgiris area was brought under four divisions called "naadus" by the badagas. The Nilgiri Mountain Railway line was built by the British (1891–1908) to access Ooty, which was later included in the UNESCO world heritage site in 2005[2, 3, 4, 5] (Breeks 1873a, b, c, d; Hockings 1980, 1998; Francis 1908).

The badagas are one of the important indigenous people, and the largest single backward class community in the Nilgiris. The badagas differ from other communities such as todas, kotas, irulas and kurumbas in the region. Badaga men and women wear a distinct attire. The economic sources of badagas are mainly agriculture and tea plantation. The major sub-groups of the badagas are: (1) haruva or brahmin; (2) athikiri or magistrate; (3) kanakka or accountant; (4) gowda or cultivator; (5) thoraiya or service-provider. Most sub-groups intermarry; marriage with non-badagas is rare and not at all encouraged. The customs governing local festivals, marriage, funerals etc. are exclusive to the badagas. The history, stories, proverbs, prayers, costume, songs, music and dances are also very special and unique (Hockings 1988; Grigg 1880; Thurston and Rangachari 1909). Research reports documented by non-badagas over centuries are available in the literature.

---

[1]https://en.wikipedia.org/wiki/The_Nilgiris_District.

[2]http://www.badaga.org/forum/viewtopic.php?t=530.

[3]https://en.wikipedia.org/wiki/John_Sullivan_(British_governor).

[4]https://en.wikipedia.org/wiki/Nilgiri_Mountain_Railway.

[5]http://whc.unesco.org/en/list/944.

Fig. 1   View of Ketti valley of the Nilgiris

Fig. 2   View of Katery waterfalls and tea garden of the Nilgiris

Because of the absence of script in the badugu language, it was believed that badagas were the indigenous native people of the Nilgiris. The history of badagas had been documented in other languages by non-badaga historians and anthropologists (mostly westerners).

This chapter focuses mainly on the unique and exclusive characteristics of badagas with special attention to the attire system that has been followed by the badagas over centuries as fashion. Irrespective of age and gender, all badagas tend to wear traditional white attire in the form of thundu, mundu, dhuppatti, seelai, mandarai, mandepattu, etc. on almost all social occasions and perform traditional badaga dances as ethnic fashion.

2 Research Reports

Edgar Thurston, British museologist and superintendent at the Madras government museum, and his colleague Rangachari[6] published seven volumes of book on "Castes and tribes of southern India" in 1909 as part of assignment towards an ethnographic survey of India. Murray Bornson Emeneau, Emeritus Professor of Linguistics and Sanskrit at the University of California, did his field study[7] in the Nilgiris in 1936. His article in the journal *Language* in 1939 on the vowels of the badaga language is still the clearest study of a unique phenomenon in the world's languages—a system of vowels that includes not one, but two degrees of retroflexion—a quality resembling the sound of English "er". Paul Hockings, Professor Emeritus at University Illinois, USA was trained in anthropology.[8] He did extensive research and field work at the Nilgiris from 1962. His publications about the Nilgiris and the badagas are: (1) a bibliography for the Nilgiri hills of southern India (1972); (2) Ancient hindu refugees—badaga social history 1550–1975 (1980); (3) Counsel from the ancients—a study of badaga proverbs prayers omens and curses (1989); (4) Blue mountains—the ethnography and biogeography of a south Indian region (1990); (5) A badaga-English dictionary (1992); (6) Blue mountains revisited (1998); (7) Kindreds of the Earth—badaga household structure and demography (1999); (8) Mortuary ritual of the badagas of southern India (2001); (9) So long a saga—four centuries of badaga social history (2013). His massive coverage of the details of the Nilgiris has paved the way for classifying him as "Nilgiriology". Commemorating the rich contributions made by Hockings, The Nilgiri Documentation Centre in Ootacamund has honoured him (in absentia) with "The Nilgiris-Lifetime Achievement Award 2015" when Prof. Hockings turned 80.[9] Frank Heidemann, Professor of Anthropology at the University of

---

[6]https://en.wikipedia.org/wiki/Edgar_Thurston.

[7]http://www.berkeley.edu/news/media/releases/2005/09/09_emeneau.shtml.

[8]http://www.uic.edu/depts/lib/specialcoll/services/lhsc/ead/011-02-20-03f.html.

[9]http://www.oneearthfoundation.in/?p=1260.

Munich, Germany, did his field work in the Nilgiris[10] focusing on religion and politics. He was the author of the book "Akka bakka: religion and dual sovereignty of the badaga in the Nilgiri south India" (2006). Christiana Pilot Raichoor, a researcher from France, worked on the constitution of corpus sound linguistic analysis of badaga oral tradition language[11] and associated with Paul Hockings in the book "A badaga-English dictionary—trends in linguistics" (1992). Claire Martin, a doctoral student of the University of California, did her Ph.D. thesis[12] on the oral epic performance tradition of the badaga tribe of the Nilgiri hills, south India.

A few reports by native badagas are also available in the literature. Jogi Gowder (1827) published an article on badaga brahmins of the Nilgiris and Belli Gowder (1923) of Ketti valley published an article on badaga funeral prayer. In recent years, the history and traditional practices are being reported by many native badaga people. Balasubramaniam, a badaga engineer from Jagathala village, published a book[13] entitled "Paame", which means "story", showcasing the unique aspects of badaga history and culture. This book was published in 2009 by Elkon Animations, Bangalore. This company was owned by another badaga engineer Raju of Yedappali village. Sivaji Raman, a badaga from Jakkanarai village, published a book[14] entitled "Baduga samudhayam", which means "Baduga community", in 2009. His book was a private publication with a foreword by Rev. Malli from Kerbetta village. Raman reports that the badaga language finds extensive mention in old Tamil literature "Tholkappiam". He added that badugu was a separate language spoken in the northern side of Tamilnadu called "Vadagam" and could be traced back 2300 years. He feared that the absence of script in the badugu language had made use of Tamil and English languages compulsory for the badagas for education, employment and communication with non-badagas. This situation had a serious effect on the badagas, leading to them forgetting certain important badugu words. Haldorai (2003a, b, 2004, 2006a, b, 2008), a badaga from Kiye Kowhatti village and a doctoral scholar in comparative linguistics, published 11 books about badagas, of which two were nominated as best books by the Government of Tamilnadu. He added that almost all the names of badaga villages were coined based on the badaga origin. Badaga children practiced using badugu as their mother tongue at their own pace; almost all were using badugu at home (with parents, children and siblings), but they used mainly Tamil and/or English as the medium of instruction at educational institutions and hence could communicate with non-badagas either in Tamil or in English. Most of the badagas read magazines and books published in Tamil or English because of the absence of

---

[10]http://www.lit-verlag.de/isbn/3-8258-8116-4.

[11]http://www.univ-paris3.fr/pilot-raichoor-christiane-29806.kjsp?RH=1179925961149.

[12]http://www.ethnomusic.ucla.edu/spring-2003-newsletter-student-profiles.

[13]http://www.amazon.com/Paam%C3%A9-History-Culture-Badagas-Nilgiris-ebook/dp/B013UIOW2A.

[14]http://badaga-books.blogspot.in.

script in the badugu language.[15] Bellie Jayaprakash, a badaga from Beratty village and a Wing Commander in the Indian Air Force, had uploaded much information[16] about the badagas in his blog.[17] Halan (1997), a badaga researcher carried out doctoral research on pastoral badagas of the Nilgiris. Ragupathy (2007), a badaga from Nanjanadu village and research scholar of Annamalai University, published Badaga–English–Tamil dictionary by providing meanings of 3500 badaga words.

3 Characteristics of Badagas

The population of badagas had increased tremendously (about 600-fold) during the last four centuries. There were a meagre 500 in the year 1603, rising to 2207 (in a census made by the British in 1812), 34,176 (census in 1901) and 104,392 (census in 1971). As of a 2011 census report, the Nilgiris had a total mixed population of 735,394, in which the badagas alone account for about 4 lakhs. The badagas speak badugu language, a dialect of the Kannanda language. The badagas are distributed in four divisions (locally known as naakku seemai). Over 400 villages (locally known as hattis) come under four seemais—Porangaadu, Dhoddanaadu, Mekkunaadu and Kundah.

It is assumed that badagas are Adi dravida by descent, Hindu by religion, Saiva by sect and nature worshipers. Even today they worship stones, with nature being a central core. This can be illustrated by the single stone, two pillar and four pillar stone structures (Fig. 3) built more than 100 years ago at Katery village.

Special pooja and prayers were conducted during the celebration of the 119th year of the Hiriodaya temple (Fig. 4) on December 7, 2015 by the badaga members of Katery village for the welfare of unknown non-badaga members who have been affected in Chennai and Cuddalore areas by the heavy rain and floods in November, 2015.

UNESCO declared in 2009 that hospitality of the badagas in the Nilgiris was the "best" and conferred a global honour.[18] Worship of stone and stone structures suggests that badagas, as were the ancient Greeks and Egyptians, of an ancient ethnic group.[19] Research reports state that the badagas might have migrated from Central or East Europe and accepted the local language for survival in the Nilgiris between 1500 and 1600 AD, which then belonged to the Vijayanagaram, hence the dialect of Kannada. Genomic studies on Y-chromosome DNA marker tests on the

---

[15]http://badaga.co/the-story-of-the-badagas-paame-by-bala.

[16]http://badaga-first.blogspot.in.

[17]http://badaga.co.

[18]http://www.thehindu.com/todays-paper/tp-national/tp-tamilnadu/badaga-hospitality/article221752.ece.

[19]http://badaga.org/forum/viewtopic.php?t=574&sid=42151d0002ea63a646b7143bfc4dc6f6.

Fig. 3 Stone structures at Katery village

Fig. 4 Hiriodaya temple built during 1897 at Katery village

badagas have also revealed that badagas belonged to the broader R1a and specifically R1a1 Haplo group of people in Europe.[20]

Badaga villages (hattis) and their houses (mane) are also very distinct and unique. Hattis are generally situated on or near the summit of a low hillock, composed of rows of tiled or terraced houses and surrounded by agricultural fields.

---

[20]http://en.wikipedia.org/wiki/Badagas.

Fig. 5 View of badaga houses in a row

The houses are not separate but built in a line under a continuous roof and divided by partition walls (Fig. 5).

It appears that, when the badagas established their villages, they were selective on two important environmental factors: (1) houses laid out in rows, each row facing east as far as possible; (2) availability of water resources nearby. It is noted that not a single village was built in low lying areas as a protective measure to avoid flood and other natural calamities. The houses had common walls, possibly on warmth and safety grounds. Every house had almost a similar inner partition structure; boilu (entrance), nadu mane (living room), somi mane (pooja room), oge mane (inner room) used as a kitchen, pille (bathroom) and ere mane (office) adjacent to boilu. The entry to the pooja room was mostly through an arch with provision for placing a holy lamp (Fig. 6).

In some houses, a loft was structured over the oge mane to serve as a storehouse. Drying and threshing operations on food grain were carried out on the ground in front of the houses, called keri. Wastewaters emanating from kitchen

Fig. 6 View of arch with provision for lamp in the middle of a house

and bathroom were disposed to the backyard of the house, called imbara. The cattle (cows and buffalos) were bought to the keri, fed with grass, provided with water and milked. In the evening, the cattle were kept in stone kraals or covered sheds (kottege) close to the houses. The litter of the cattle was carried away as manure for the agricultural lands and tea estates.[21]

Badagas worship ancestral Goddess "Hethey" and ancestral God "Hiriodaya". Habba (means festival) system is unique. The chief festival of the badagas is "Hethey habba". It has been celebrated with divine power and traditional gaiety usually in the months of December and January each year at different villages (hattis) such as Bandhimai, Bebbanu, Beragani, Chinna Coonoor, Ebbanaadu, Gasu gui, Hallukeru, Honnadhalai, Horanalli, Jegathala, Kakkarayar bettu, Kanneri, Ketti, Kookkal, Nattukkal, Nunthala and Pedhuva. Photography of Hethey idols or Hethey temples has been strictly prohibited in the majority of Hethey villages. The Hethey habba practiced at Beragani village has been very special. It is said that a non-badaga weaver comes with his portable handloom and sufficient thread for weaving coarse cloth and turbans. There is a special house, in which these articles are woven. At other villages where the Hethey festival is observed, the badagas go to the weaver's place to fetch the required thread or cloth. After bathing early in the morning before the festival, a few older badagas and the weaver proceed to the weaving house. The weaver sets up his loom and worships it by offering incense and others. Badagas give him new cloth and a small sum of money, and ask him to weave a coarse cloth (dhuppatti) and two narrow strips of cloth (thundu). On the morning of the last day, the poojari,

---

[21]http://badaga.co/badaga-villages-nakku-betta-hattis.

accompanied by all badagas, takes the newly woven cloth to a stream, in which it is washed. When it is dried, all members proceed to the temple, where the idol is dressed up and decorated with waist and neck ornaments and an umbrella. Thereafter, devotees are allowed to have darshan. The devotees bring round coins (kanikke)—specifically tied within small pieces of white cloth and brought from their houses—as small offering and place them on a tray near the idol. In a similar manner, all other Hethey villages also celebrate the Hethey festivals at their temples at different dates during December and January each year. Lakhs of devotees all wearing white dresses visit the Hethey temple in a particular village on the festival day, submit their prayers, perform dances and have a free meal. It should be mentioned that the provision of meals to almost over one lakh devotees on the festival day necessitates a very special system by badagas. Entire ingredients (rice, vegetables, groceries, milk etc.) required for preparing the meal are donated by the badagas themselves as charity; preparation and serving of meal also is done by badaga volunteers as a social service. The author, Gurumallesh Prabu, attended both the God Kariabetrayar festival at Nellithorai and the Goddess Hethey festivals held at Beragani, Jegathala, Ketti and Nunthala (Fig. 7) and took photographs of the devotees from outside the temple area.

It is customary that every male devotee should wear white dress when attending the Hethey festival. Figure 8 depicts the image of the author Gurumallesh Prabu with his father-in-law on the way to the Beragani Hethey festival during 2007.

Specific thing to mention here is that almost all younger male devotees wear only a white shirt (kamis) and white dhoti (mundu), whereas older male devotees wear a white shirt, white dhoti, white shawl (seeley) and white turban (mandarai) during Hethey festivals. Older female devotees wear all white thundu, mundu and

Fig. 7 Kariabetrayar festival at Nellithorai and Hethey festivals at Jegathla, Ketti and Nunthala

Fig. 8 Badaga men's attire

mandarai pattu combination attire, whereas younger female devotees wear a saree and blouse combination with an additional white mundu wrapped over. All white attire is considered as divine and pure.

In the badaga community, the position of the women is very peculiar. Almost every badaga has a few acres of land to cultivate. Women do all the farm work and men engage in earning something in hard cash. To a badaga man, his wife is his capital. A badaga woman who does not have land of her own to cultivate finds work on some other badaga land. Women work hard for their husband and family. In a typical case, the grandmother of the author, Dhali (Fig. 9), was looking after her agricultural farm work both at Manihatty village and at Katery village and his grandfather, Belliah, was working in a cordite factory, Aruvankadu during the middle of the last century in the Nilgiris. The attire system adopted during the 1940s is being followed by elders even in 2015; a relative from Jegathala village wearing the traditional attire can also be found.

Generally, badaga women are held in high esteem. This could be for three main reasons: (1) absence of a dowry system in marriages; (2) divorce by mutual consent; (3) widow re-marriage. There is no stigma attached to widows. They are also

Fig. 9 Badaga women's attire—past and present

respected as part of the mainstream and brought to the forefront in functions such as engagements and weddings.

Wearing of a turban is considered to be a mark of stage, dignity and pride of elders. Whether it is a marriage or a funeral, most middle aged and older badaga men wear kamis, mundu, seelai, turban and gather together to oversee and to take part in conducting the ceremonies (Fig. 10). Badaga women wear a white body-cloth (mundu), a white under-cloth tied round the chest (thundu) and a white mini cloth worn as a cap (mandepattu) (Fig. 11).

The distinctive attire of badaga women was said to be the disguise adopted by them in flight. Tattooing on the forehead of teenage girls was the sign of avail-ability of that girl for marriage. The tattoos on foreheads and forearms, a meas-ure taken to make them unattractive in an earlier period, could only be seen on a few older women in recent times. To the best of the author's knowledge, tattooing almost vanished as their civilisation progressed. The fashion and passion for wear-ing all white thundu, mundu and mandepattu is a decreasing trend because of the introduction and adoption of saree-blouse styles by middle-aged women and the interest in the half-saree or tops-chudhithar style by teenage girls.

The badaga marriage system is very simple and unique. Most marriages are of the arranged type only, and love marriages are few. It is a custom that the bridegroom's family visits the residence of the bride on the day before the mar-riage, stay there and perform the custom of buying the bride by paying rupees 200 as "onnu" (in the majority of cases). From then onwards, the bride would be the member of the bridegroom's family. On the day of the marriage the bride

Fig. 10 Older men of Thangadu village

Fig. 11 Older women of Kappachi village

Fig. 12 Marriage of Gurumallesh Prabu with Vijaya

(normally from another village) is brought to the bridegroom and the marriage is conducted either in the house of the bridegroom or in the temple premises. Most villages have now built community halls for conducting marriages and other functions. The marriage of the author Gurumallesh Prabu (of Katery village) with Vijaya (of Kollimalai Horanally village) was conducted at Katery Murugan temple in 1988 (Fig. 12). A simple decoration (pandal) was made in front of their house. It should be noted that his parents and family members were dressed up with simple thundu, mundu, mandarai style on the day of the marriage (Fig. 13).

Similarly, the marriage of his sister Geetha with Belliraj Mohan of Kammandu village was conducted at Kammandu village (Fig. 14).

It should be noted that great importance was given to the badaga marriage custom rather than to the dress or jewellery. In these two cases of marriage, all the villages (Katery, Horanally and Kammandu) of brides and bridegrooms come under the same Mekkunaadu seemai. The marriage of his daughter Poorani (of Mekkunaadu seemai with haruva sub-class as pure vegetarian) with Sasikumar (of Jakkanarai village in Porangaadu seemai with baduga sub-class as non-vegetarian) was held during 2014. The marriage ceremony began with the formal consent of

Fig. 13 Members of the family on the day of the marriage

Fig. 14 Marriage of Geetha

Fig. 15 Obtaining consent of the bride

Fig. 16 Event of Onnu kattodhu

the bride (Fig. 15) and then the buying of the bride (onnu kattodhu) performed in
the house of the bride at Katery village (Fig. 16). The marriage ceremony was held
at a hotel in Ooty town, the Queen of Hill Stations (Fig. 17).

Similar badaga marriages were conducted in the authors family in recent
times: (1) Aarthi of Katery village got married to Jeganathan of Sogathorai village

Fig. 17 Marriage of Poorani with Sasikumar

(Fig. 18) in a function at Coonoor town; (2) Remadevi of Kappachi village got married to Arunkumar of Kammandu village (Fig. 19) in a function held at the residence of bridegroom at Kammandu village; (3) Supriya of Thuneri village got married to Sanjey of Jakkanarai village (Fig. 20) in a function held at Kotagiri town.

In all these recent marriages, it was observed that the costume of the bridegroom was as usual (all white shirt, dhoti, seelai and turban), although much attention was paid to get the latest and modern costume and jewellery for the bride. In these marriages, intermarriage was adopted between the badaga sub-groups under same seemai as well as with other seemai.

The funeral of Sarasu, mother of Gurumallesh Prabu, took place at Katery village in 2013 as per the badaga custom. At her residence, the body (Fig. 21) was initially covered with thundu, mundu, seelai and mandepattu, and then it was brought to the neighbouring street (keri) and kept within a temporary metal frame structure decorated with coloured cloths, flowers, balloons and an umbrella at the top. Immediate family female members (both young and old) sit around the body (Fig. 22) and others form a gathering nearby with their heads covered by either dhoti or shawl (Fig. 23).

Fig. 18 Marriage of Aarthi with Jeganathan

Fig. 19 Marriage of Arunkumar with Remadevi

Fig. 20 Marriage of Supriya

Apart from the unique social and voluntary systems adopted by badagas during marriages and festivals, another speciality system to mention here is the practice of badagas in the death and funeral system. Every male member of a family in a village pays a monthly subscription towards the death fund to a village person assigned for this purpose. That person collects the money and keeps it in reserve. When a death occurs, he distributes a fixed lump sum to the deceased family to meet the funeral expenses. All funeral activities (spreading word of the death, decorating the corpse frame structure, preparing and serving of a meal, burial at graveyard, etc.) are taken complete care of by the social youth volunteers of that village under the leadership of the headman. This system is being adopted in almost all badaga villages in the Nilgiris.

Badaga music and dances are captivating. There won't be any function or ritual without the traditional badaga group dance. Elders still follow the traditional badaga dance (Fig. 24) but youngsters have modified the dancing style to be more attractive to meet current fashion.

Fig. 21 Body covered with seelai

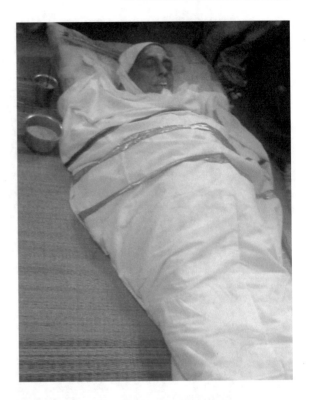

Dancing is an integral part of the badaga system as is white attire as fashion. At a special event in August 2015, a group of about 600 badaga devotees performed the 51st pathayatra (devotional bare-foot walking event) from the lord Muruga temple of Katery village in the Nilgiris to the lord Muruga temple in the Palani hills, covering a distance of about 200 km on roads in bare feet in a total period of 5 days wearing devotional attire. The devotees, family, friends and relatives performed a massive badaga group dance at the temple shrine on the last day (Fig. 25). It is a custom that, in the badaga rituals and functions, the closing ceremony is performed with chanting "Havu Ko, Hey Havu Ko" repeatedly and collectively by all, and by the majority of members in white attire.

Whenever a badaga person meets another elder or aged badaga person of repute, the younger person seeks blessings by bowing the head. In turn, the elder or aged person places his or her hands on top of the head of the youngster with both palms open and blesses by saying "let everything become good, let one become thousand (wealth), let KO be the call, let it boil as BO, let God give good health and happiness, let only good things happen while you go out or come in, let the education take you forward, let you take all that is good, let you leave all that is bad, let you be a great person, let you lead the nation, let you live over 100 years, let you visit all over the nations, and let you come back safely with winning face". The blessings are long-term advice from older badagas to the

Fig. 22 Funeral: tribute point

Fig. 23 Section of women during funeral part

Fig. 24 Badaga group dances during festival, marriage and funeral

Fig. 25 Pathayatra group dance at Palani Temple

younger badaga generations. The elders expect the youngsters to grow in stages; from the place of the Nilgiris as a local point to national and international places as target points, to preserve the custom and values at their places of stay, and to return to the Nilgiris with enriched knowledge and power.

Above are a few examples of the unique characteristics of the badagas. These reveal the exclusive system of badagas on social and cultural fronts. In all major events, such as marriages, festivals and funerals, the attire system (all white dress) was invariably adopted by most. In recent times, new systems and technologies

Fig. 26 Badugu language script

have been introduced to sustain and to document the badaga cultural and custom practices. Since 1989, "Badaga Day" has been celebrated on May 15 every year by the badagas to showcase their solidarity.[22] It is also celebrated as "Ari Gowder Day", commemorating the contributions rendered by Rao Bahadur Ari Gowder of Hubbathalai village for the badaga community.[23] For the development of a script to the badugu language, Anandhan Raju designed badaga script based on the shapes of Tamil characters during 2009. In 2012, Yogesh Raj of Kadasolai village released script for the badugu language (Fig. 26) and has been training the badagas to learn badugu. Saravanan Raju and others also contributed to the development of badugu script.[24]

For the dissemination of local information through electronic media, "Nakkubetta TV" channel was established in 2013. It runs under Nakkubetta Foundation and covers the districts of the Nilgiris and Coimbatore, nurturing the traditional and cultural values of badagas. Nakku (means four) Betta (means mountain) has been coined based on the existence of four green mountains (also four seemais) in the Nilgiris. The umbrella in the logo is a symbol of care, shelter, royalty and protection of badaga tradition. This TV channel has been advanced to the next stage as WebTV to reach all around the globe.[25] Badaga local news bulletin "Seemai sudhdhi" has been on air since 2013 from the Ooty main radio channel MW 1602 as well as FM 101.8. For the development of activities through social media networks, "Badagas!" group was created in the Facebook portal. This

---

[22]http://badaga-day.blogspot.in.

[23]http://timesofindia.indiatimes.com/city/coimbatore/Badagas-to-dig-into-their-past/articleshow/13159221.cms.

[24]http://scriptsource.org/cms/scripts/page.php?item_id=entry_detail&uid=w5qdq83f7s.

[25]http://www.nakkubetta.tv.

closed group is administered by the badaga members of the Nilgiris. As of December 31, 2015 the Facebook account shows over 14,000 badaga members in this group. Many badaga members post and share the cultural, social, environmental and similar information in this Facebook group. This is found to be very useful for badagas of all age groups all over the world to understand and update details about the Nilgiris and the badagas.[26] International badaga association[27] is also providing information through its webpage for the welfare of badagas across the globe. Movies and websites have been created by the badagas for entertainment. Large numbers of badugu songs have been added in dedicated websites for online listening for music lovers.[28] Badaga associations and individuals have released many cassettes and compact discs containing badaga audio and video songs. The first badaga film "Kaala thappitha payilu" was produced in 1979. Films such as kemmanju, hosa mungaru, gavava thedi and sinnatha boomi have since been produced.[29]

It is strongly hoped that every badaga living both within and outside the Nilgiris has a sense of belonging in the protection and sustenance of their unique and rich culture. The badagas strive hard to preserve and promote their culture in and out of the Nilgiris. Wearing of white dress (as attire) on almost all customs and rituals is considered a traditional ethnic fashion.

4 Conclusion

Badagas were little known outside the Nilgiris earlier. Now, badagas live in other cities in India and abroad in sizable numbers for various reasons. Today, badagas own a substantial share of tea plantations in the Nilgiris and almost monopolise the professions of doctors, lawyers, engineers and teachers within the Nilgiris and a growing number outside the Nilgiris. Even then, about 10 % are affluent, 50 % are middle class and 40 % are living subsistence at the subsistence level. Most badaga sub-groups intermarry, but marriage with non-badagas is quite rare although the number has been rising in recent years. Even though migrated badagas try to follow their custom and culture outside the Nilgiris, it is feared that the culture is slowly losing its shine. The badaga attire referred to here is mainly worn only at ceremonial functions and events. Not all badagas follow such practices all the year round. If this trend continues, badagas would lose their identity in the years to come. Badaga ancestors had their own house, land and food. As civilization progressed, a few badagas have lost their food system and lifestyle but badagas are rich in hospitality and customs. Moreover, research reports or other

---

[26]https://www.facebook.com/search/str/badagas/keywords_top.

[27]http://www.badaga.org.

[28]http://www.baduguraaga.com.

[29]https://en.wikipedia.org/wiki/Badaga_cinema.

inputs are being submitted or shared by the badagas through ICT and social media at large. It is hoped that these modern tools can support the process of preserving and promoting the unique cultural systems of badagas. The system of attire (white dress) has not been changed much and this can be considered as a fashion and local strength.

Acknowledgments   The authors wish to thank all their family and village members for their kind permission to publish the details in this chapter. All images depicted are original and owned by the authors.

References

Belli Gowder MK (1923) Hill tribes social reform. Private Print, Ootacamund

Breeks JW (1873a) An account of the primitive tribes and monuments of the Nilgiris. India Museum, London

Breeks JW (1873b) Nilgiri Manual 1:218–228

Breeks JW (1873c) Madras J Sci Lit 7:103–105

Breeks JW (1873d) Madras Mus Bull 2(1):1–7

Francis W (1908) Madras district gazetteers, the Nilgiris. Logos Press, New Delhi

Grigg HB (1880) A manual of the Nilgiri District in the Madras Presidency. Government Press, Madras

Halan J (1997) Pastoral badagas of the Nilgiris-a case study. Ph.D., thesis of Bharathiar University

Haldorai RK (2003a) Badaga proverbs. Nelikolu Publishing House, Udhagamandalam

Haldorai RK (2003b) Hethe Deivam. Nelikolu Publishing House, Udhagamandalam

Haldorai RK (2004) Goddess Hethe (Goddess Hethe). Nelikolu Publishing House, Udhagamandalam (Tamil version)

Haldorai RK (2006a) Badugu—a dravidian Language. Nelikolu Publishing House, Udhagamandalam (Tamil version)

Haldorai RK (2006b) Marriage among the Nilgiri Badagas. Nelikolu Publishing House, Udhagamandalam (Tamil version)

Haldorai RK (2008) Dewa festival of the Badagas. Nelikolu Publishing House, Udhagamandalam

Hockings P (1980) Ancient Hindu refugees. Studies in Anthropology, Book VI, De Gruyter Mouton

Hockings P (1988) Blue mountains: the ethnography and biogeography of a South Indian region. OUP, India

Hockings P (1998) Blue mountain revisited—cultural studies on the Nilgiri hills. Oxford University Press, Oxford

Jogi Gowder BK (1827) The origin of the Badaga Brahmins of Nilgiris, and the Temple of "Mahalinga" at Melur village in Merkunaad Division in the District of Nilgiri. Private Print, Ootacamund

Ragupathy R (2007) A Badaga–English–Tamil dictionary. Diva Bharathi Publications, Ooty

Thurston E, Rangachari K (1909) Castes and tribes of Southern India, vol 1–7. Government Press, Madras

Authors Biography

Gurumallesh Prabu a native badaga of the Nilgiris, has been serving as Professor of Chemistry at Alagappa University, Karaikudi, India since 1987.

Poorani daughter of Gurumallesh Prabu, is a biotechnology doctoral research scholar at Kumaraguru College of Technology, Coimbatore, India. She was a recipient of DST-KVPY Government of India National fellowship award for 4 years (2010–2014).

Ethnic Styles and Their Local Strengths

Nithyaprakash Venkatasamy and Thilak Vadicherla

Abstract Ethnic style characterizes a culture which celebrates heritage and origin and has evolved over a period of time. Ethnic style is influenced by factors such as availability of a wide variety of materials, production methods, popular designers, organizations, cross cultural influences, innovations, and sustainable practices. Ethnic styles have evolved geographically around Asia and the Pacific, Africa, Latin and Central America, North America, and Europe. Acculturation has helped in not only westernization of clothing but also produces cultural diversity and heterogeneity. Availability of sustainable raw materials (natural fibers, natural dyes, recycled materials) combined with sustainable manufacturing/chemical processing/surface embellishments (use of resist bases, hand painting, itkat, batik, bandhi, natural dyes, and effluent treatments) have added value to ethnic styles. Artisanality combined with the creativity of the designers and involvement of organizations such as Abury champions have created a platform for the evolution of sustainable ethnic styles. Generational changes (Generation X and Generation Z) can be seen as a positive influence and ethnic styles avail themselves a great chance to co-exist with modern fashion concepts.

Keywords Acculturation · Artisanality · Designers · Ethnic   styles · Evolution · Generation Z · Organizations · Sustainability

N. Venkatasamy (✉) · T. Vadicherla
Department of Fashion Technology, Bannari Amman Institute of Technology,
Sathyamangalam, Erode, Tamilnadu, India
e-mail: nithyaprakashv@bitsathy.ac.in

T. Vadicherla
e-mail: thilak.vadicherla@bitsathy.ac.in

© Springer Science+Business Media Singapore 2016
M.A. Gardetti and S.S. Muthu (eds.), *Ethnic Fashion*, Environmental Footprints
and Eco-design of Products and Processes, DOI 10.1007/978-981-10-0765-1_6

1 Introduction

Since the 1990s there has been an increasing awareness of environmental and cultural ecological protection, culminating in ethnic concepts adoption for fashion garments (Li 2009). The ethnic capital was broadly zoned into from Asia (China, India, Japan, Middle East, etc.), Europe (traditional folk costumes of the Nordic and the Eastern European), South America (Indiana, Mexico, etc.), and Africa. Ethnic dress can be defined as "the clothing related to a specific ethnic or cultural group sharing the same heritage, background, and beliefs, relaying a symbolic message to others allowing group categorization by distinguishing a group from another by differentiation" (Eicher 1995; Eicher and Sumberg 1992).

For example, an American Indian endeavors to strike a balance between the Indian ethnic identity and American fashion by sporting jeans, T-shirt, nose ring, Punjabi style braided hair, and a bindi (Sunita 2001). The youngster is also mindful of the fact that the combination does not strike a direct confrontation with the western fashion look, but it reflects the attitude of the current generation youngsters who try to strike a balance between appearing fashionable and at the same time sticking to cultural roots. In multicultural environments, clothing style invites multiple interpretations among different sections of people (Sunita 2001). The Indian diaspora would interpret it as self confidence whereas American nationals would interpret it as modesty, sophistication, etc.

Clothing style is influenced by socio-psychological factors, cultural factors and, demographic profile (Sunita 2001). Fashion is not only innovative in producing new looks—it also witnesses diffusion of cultural practices. A cultural symbol in one country might end up as a fashion statement in other. Roland Barthes's lexicon fashion garment is an amalgamation of several dress separates with a certain number of implicit variants such as form, movement, stiffness, identity, etc. which constitutes a syntagm comprising several elements. The fashion theorists consider it as a case of symbolic interactionism, where every individual makes decisions and forms opinions from their own interpretation and evaluation. Thus in this post modern consumerist society, every clothing style has an equal chance of being combined along with other elements to constitute a fashion statement. A glance into some of the enterprising and resourceful ethnic styles across the globe reveals an understanding of the ethnic styles but time alone indicates which style is adapted and constituted as fashion clothing.

2 Evolution of Ethnic Styles

2.1 Review of Ethnic Dresses Throughout the World

2.1.1 Asia and Pacific

The land of Asia stretches from the dusty desert terrains of Iran in the west to the mountain ranges of Japan in the east, featuring different types of ethnic communities. Asians are well known not only for their cultural heritage but also for their

Fig. 1 Tunic decorated with embroidery or woven tapes (*Source* https://commons.wikimedia.org /wiki/File:Stans08-218_%283134170153%29.jpg)

indigenous dress traditions. Generally, Asia is zoned into four regions: Central Asia, East Asia, South East Asia, and Asia Pacific with reference to their geographical proximity and ethnography.

The dressing traditions of Central Asia feature a tunic and salwar decorated with embroidery or woven tapes. The salient feature of the tunic (Fig. 1) is that it was usually made from narrow width cotton or silk. A coat is worn above the tunic (Central Asian Women's Adornment and Clothing 2014) in the winter season. The salwar also is richly embroidered at the cuffs and peeps out from below the kurta. The overcoat khalat is more elaborate, comprising stripes and rich ikat patterns. The finest of all this clothing is the velvet tape used for embellishing the overcoat (Valerie 2005c).

East Asia is culturally dominated by China. Among all the Chinese costumes, cheongsam and mao suit have prevailed and have undergone a lot of style changes.[1] Figure 2a represents a typical cheongsam, which is a fitted, one-piece women's garment with mandarin collar and side-slit skirt. A cheongsam is generally made of silk brocade, satin, satin brocade, or velour (Valerie 2005d). These fabrics are laced with symbols of dragons, floral shapes, and Chinese icons.

Figure 2b represents a typical Mao suit that comprises a front-fastening jacket and trousers of the same material but it is rarely used by contemporary Chinese men except for special occasions (see Footnote 1).

The kimono (Fig. 3) is one of the most sought after garments among Japanese. Men and women wear kimonos on special occasions such as weddings, tea

---

[1]http://www.my-qipao.com/china_attire_e1.html. Accessed 29 November 2015

(a)   (b)

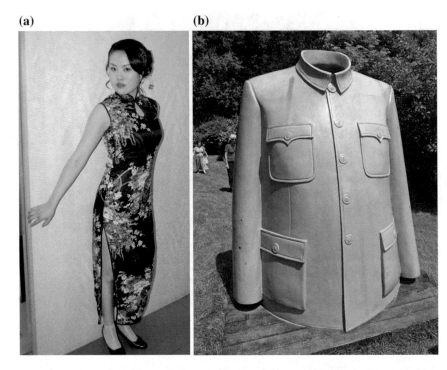

Fig. 2 a Cheongsam (*Source* https://commons.wikimedia.org/wiki/File:Chinese_woman_in_Qipao.jpg). **b** Mao suit (*Source* https://commons.wikimedia.org/wiki/File:Mao's_coat_-_geograph.org.uk_-_854641.jpg)

ceremonies, and formal gatherings. Kimonos come in colorful and mesmerizing patterns, although kimonos dyed over the whole surface are preferred for formal occasions.[2] Unlike wholly wrapped styles, kimono dress is stitched from panels and wrapped around the body. Notwithstanding the origin of the kimono in Japan, it also exists in Korea in a different variant form known as a hanbok (Fig. 4) (DeLong and Key Sook 2009; Valerie 2005e). Indonesia is known for its batik printed shirts and sarongs (Fig. 5a, b).[3] Traditional sarongs are basically ikat weave fabrics (Valerie 2005g). Figure 6 represents the national costume of Vietnam[4] (Valerie 2005c) Ao dai, literally meaning long shirt, is a long tunic with a close fitting bodice, side slits, and raglan sleeves.

South East Asia is home to salwar kameez, ghagra/lehenga choli, pajamas, pirans, mekla chaddar, sarees, and dhoties (Indian Traditional Costume and Makeup 2010) which are typically represented in Fig. 7a–f. Because of acculturation, the vast variety of styles has evolved a great deal ever since their early

[2]https://en.wikipedia.org/wiki/Kimono. Accessed 09 November 2015.

[3]https://en.wikipedia.org/wiki/National_costume_of_Indonesia. Accessed 15 November 2015.

[4]https://en.wikipedia.org/wiki/Ao_dai. Accessed 21 November 2015.

Fig. 3 Kimono (*Source* https://commons.wikimedia.org/wiki/File:Hall_of_Japan_04.jpg)

Fig. 4 Hanbok (*Source* https://commons.wikimedia.org/wiki/File:Korean.clothes-Hanbok-01.jpg)

(a)

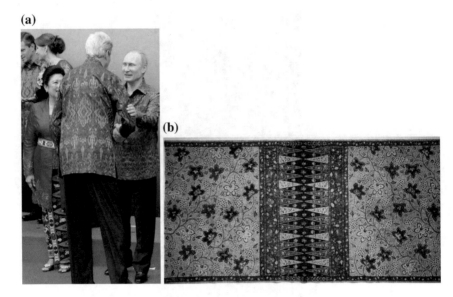

(b)

Fig. 5 **a** Batik printed shirt (*Source* https://commons.wikimedia.org/wiki/file: Secretary_Kerry_ Meets_Russian_President_Putin_at_APEC.jpg). **b** Sarong (*Source* https://commons.wikime dia.org/wiki/File:Sarong,_Northern_Java,_Indonesia,_1900-1910,_cotton_-_Cincinnati_Art_ Museum_-_DSC04349.jpg)

existence to suit today's needs. Ethnic styles are still seen in ordinary domestic life, wedding ceremonies, festivals, and family functions (Indian Traditional Costume and Makeup 2010), and even in corporate clothing Indian ethnic patterns have been used. Gho is an elegance personified, a simplistic knee length gown worn by Bhutanese men (Valerie 2005f). The wrapping style and silk scarf form the hallmark of this amazing outfit (Fig. 8).

The most emboldened and free spirited of all Asian regions is the Pacific islands. Pareo or pareu is a wrap around skirt available in different dimensions (Fig. 9) in this region and this is considered as the best way of dressing by the beach side (Valerie 2005h).[5]

2.1.2 Africa

Kenyan kitenge and dashiki need no mention in the African region. Kitenge is very popular not only in Kenya but also in many other African countries.[6] Kitenge (Fig. 10), a wrapped garment, is worn in ordinary life, ceremonies, and

---

[5]https://en.wikipedia.org/wiki/Pareo. Accessed 25 November 2015

[6]http://www.tessworlddesigns.com/products/copy-of-african-print-fabric-mat. Accessed 28 November 2015.

Fig. 6 Vietnam's Ao dai (*Source* https://commo ns.wikimedia.org/wiki/ File:Junolino_hue_2.jpg)

non-official events (Valerie 2005a). Dashiki (Fig. 11) is adopted by westerners and Europeans alike in their own style, for example dashiki shirts are combined with dress trousers, etc.[7] African black identity and black nationalism are expressed by the wearing of African and African-inspired dress such as the dashiki (Valerie 2005i). Moroccans are overall a stylish and smart nation and they prefer to dress up with fashionable and impressive clothing. Women wear caftans embellished with rich embroidery at the garment closures.[8] Generally the garments are heavily adorned and fancily stitched. Among Moroccan kaftans, the most popular is the beach kaftan (Fig. 12), a sort of beachwear that is stylish yet pretty comfortable. Abacos, kanga, batik, kente cloth, and buba are some of the other cultural products of African diaspora (see Footnote 8). Figures 13a, b and 14 depict the typical kente, kanga, and buba garments. Another distinctive silhouette featured by the

---

[7]https://en.wikipedia.org/wiki/Dashiki. Accessed 09 November 2015.

[8]http://readafrica.org/article/WHEBN0000746858/Kaftan. Accessed 29 November 2015.

(a)   **(b)**   **(c)**

(d)   **(e)**   **(f)**

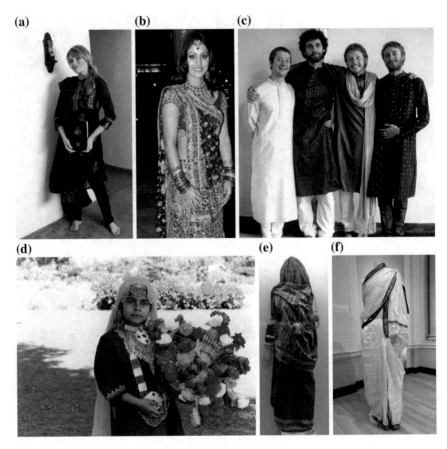

Fig. 7 a Salwar kameez (*Source* https://commons.wikimedia.org/wiki/File:Girl_in_salwar_k ameez.jpg). **b** Ghagra/lehenga choli (*Source* https://commons.wikimedia.org/wiki/File:Hina_ khan_1.jpg). **c** Pajamas (*Source* https://commons.wikimedia.org/wiki/File:Kurtas_Indian_ Dress_Different_Colors.jpg). **d** Pirans (*Source* https://commons.wikimedia.org/wiki/File: Bihar_girl_in_Kashmiri_dress.jpg). **e** *Sarees* (*Source* https://commons.wikimedia.org/wiki/ File:Sari_draping_style_of_Sambalpur_region,_Odisha_state_museum,_Bhubaneswar_2013. jpg). **f** Dhoties (*Source* https://commons.wikimedia.org/wiki/File:INDIA_2014,_polyester-silk_ blend_-_Saint_Ignatius_Church,_San_Francisco,_CA_-_DSC02646.jpg)

Namibians is a floor length skirt and fitted blouse top. It is an adaptation of Victorian dress and its masai beadwork is one of the most recognizable and reputed beadwork styles (Valerie 2005a).

2.1.3 Latin and Central America

Chilean women wear a chamanto (Fig. 15), a poncho-like garment that is warm and attractive. It is made from wool, cotton, and even silk (Traditional Costume

Fig. 8 Gho of Bhutan (*Source* https://commons.wikimedia.org/wiki/File:Bhutan_-_Flickr_-
babasteve(31).jpg)

Fig. 9 Pareo or pareu
(*Source* http://malinda4bray4.
wikidot.com/blog:212)

Fig. 10 Kenyan kitenge
(*Source* https://commons.wi
kimedia.org/wiki/File:Binti_
wa_kiafrica.jpg)

Fig. 11 Dashiki (*Source*
https://commons.wikimedia
.org/wiki/category: images_
from_wiki_loves_Africa.jpg)

of Chile 2015). Figure 16 shows the typical Ecuadorian women's traditional,
yet contemporary clothing that includes wide-brimmed woven hats with mark-
ings indicating ethnicity and pinned anakus, or wrap-around skirts and ilikllas,

Fig. 12 Beach kaftan
(*Source* https://commons.w
ikimedia.org/wiki/category
:COLLECTIE_TROPENM
USEUM_Versierde_bruidsj
urk_van_brokaatweefsel_T
Mnr_3782-817)

or mantles (Valerie 2005b). The additional adorning elements of these dresses
are decorative earrings and necklaces. The Guatemalans (Fig. 17) feature a distinctive style marked by red and white striped trousers, embroidered shirts and a
straw hat. The ponchos and Inca dress constitute the ethnic dress of Peru (Peruvian
National Costume 2015). Inca (Fig. 18) is the most decorated of all with French
and Japanese lace combinations.

2.1.4 North America

Mexican costumes (Fig. 19) are very bright, beautiful, and characteristic of sun
protection. The most popular and well-known men's pieces of clothing in Mexico
are sarape, charro suit, sombrero, guayabero, baja jacket, and poncho (Valerie
2005b). Women's dress features different variants of skirts and tunics referred
to as huipil, quechquémitl, and rebozo (Valerie 2005b). The most dazzling and
elegant designs were undoubtedly those of the Haida from the Queen Charlotte
Islands off the coast of present-day British Columbia in Canada. Totemic art
embodied in monumental totem poles and decorated house villages also mark
themes in their clothing. Large rain hats, caps, various forms of capes and wraps,
dresses, kilts, leggings, and even shoes fit the needs of autumn and winter (Philip
2005a). Moccasins made from tanned deer, moose, or caribou hide require no
mention.

(a)   **(b)**

Fig. 13 a Kente. (*Source* https://commons.wikimedia.org/wiki/category: Man_dressed_ in_Kente.jpg). **b** Kanga (*Source* https://commons.wikimedia.org/wiki/file: Brides_maids_in_ leaf_print_kanga_wraps.jpg)

2.1.5 Europe

Eastern Europe is home to rich folk costumes comprising skirts, white blouses and vests; shawls with decorative borders, and headgear. Men wear jackets, overcoats, fitted pants, and knee-high boots decorated with oriental patterns in embroidery or braid. Even the boot tops are decorated (Philip 2005a). Women's dress, which depends heavily on Western fashion, features Ottoman-influenced tulips and carnations (Philip 2005a). In Romania and Moldavia, embroidered linen blouses are among the most treasured components of folk dresses (Valerie 2005k). In Germany, Switzerland, and Austria, women's costumes feature dirndl skirts, fitted vests, white blouses, and distinctive caps, and men's include colorful vests and coats worn with knee breeches or leather hose.[9]

---

[9]http://www.thelovelyplanet.net/traditional-dress-of-germany-the-identical-emblem-of-germanic-peoples. Accessed 29 November 2015

Fig. 14 Buba tops and Iro (*Source* https://comm ons.wikimedia.org/wiki/ file:BUBATOPAIROD_ African_Lace_VLM_31.jpg)

Fig. 15 Chamanto (*Source* www.backstrapweaving.word press.com)

Fig. 16 Ecuadorian women's dress (*Source* https://zellnertravel2013.files.wordpress.com/2013/02/img_0635.jpg)

Fig. 17 Guatemalan dress (*Source* https://janewright2guatemala.file.wordpress.com/2012/11/img_0707.jpg)

Fig. 18 Peru's ethnic dress (*Source* https://pictagram.info/gallery;dwn-inca-women-art.html.com)

Fig. 19 Mexican men's and women's wear (*Source* https://commons.wikimedia.org/wiki/file:Ja rabe_Mixteco.jpg)

Fig. 20 Portuguese folk dress (*Source* https://commons.wikimedia.org/wiki/file:Viana_do_Cast clo_Romaria.jpg)

Portuguese folk dress (Fig. 20) is characterized by fine embroidery on linen garments. Spanish Flamenco dress (Fig. 21) draws heavily from the gypsy culture. The dress features a cascade of ruffles with large ease so that dancer can swish and flick the tail of the garment during a dance (Philip 2005b; Valerie 2005l). In Holland a typical man's outfit consists of blue woolen trousers and jacket worn with a flat cap and wooden shoes (Philip 2005c). Winged white caps are characteristic of women's dress. In Lebanon, known as the gem of the Middle East, the traditional women's dress comprises colorful skirts and men wear a traditional serwal or sherwal known as baggy trousers (Fig. 22).[10] Kroje (Fig. 23) is a traditional Czech costume which normally consists of skirt, blouse, vest, apron, and hat for women (Carmen 2002). Similar to other Czech costumes, Kroje dresses are also heavily embroidered. Men wear embroidered trousers, shirts, vests, and hats.

2.1.6 Meanings of Ethnic Dress

In general, ethnic dress speaks of the artistic tastes and cultural heritage of the respective country. In other words, each ethnic dress is a glaring example of a country's coveted values cherished and revered in their society. For example, in China, the national costume cheongsam represents modesty, softness, and beauty.[11] In India, the sari is symbolic of their nation's pride and sophistication. Sari depicts the Motherhood and cultural intricacies of India. Typically, the decorations in the ethnic dress speak loudly about the enrichment methods used to

---

[10]http://www.thelovelyplanet.net/traditional-dress-of-lebanon. Accessed 30 November 2015.

[11]http://www.metmuseum.org/toah/hd/orie/hd_orie.htm. Accessed 27 November 2015.

Fig. 21 Spanish Flamenco dress (*Source* https://commons.wikimedia.org/wiki/file:Flamenco_dancer_3467.jpg)

Fig. 22 Lebanese sherwal (*Source* https://collections.artsmia.org/art/108624/sherwal-turkey-or-kurdistan)

Fig. 23 Traditional Czech
dress (*Source* https://com
mons.wikimedia.org/wiki/
file:Zandní_část_ženského_n
edachlebského_kroje.jpg)

fashion up the intended look (Valerie 2005m). Ethnic dress is a manifestation of the social traits that draw on such elements as gender, taste, ethnicity, sexuality, sense of belonging to a social group, etc., whereas fashion clothing draws upon a combination of individual values, traits, and group behavioural characteristics.

However, ethnic style is comparative to other artistic styles, such as classical style, urban style, etc. The term "ethnic style of dress" refers to modern clothes with the characteristics of a national dress style (Li 2009). These dresses are also modern because the ethnic dress style has evolved a great deal right from the inception stage in the ancient period—the manner in which it is draped, materials used, and the co-coordinating accessories. Another important feature of ethnic style is that, being artistic in nature, it avails itself a chance to serve as a standalone inspiration for fashion clothing or used in parts along with fashion clothing. A classic example is the concept of orientalism in fashion. Orientalism generally refers to the appropriation by Western designers of exotic stylistic conventions from diverse cultures spanning the Asian continent (Valerie 2005o).

Ethnic styles in different countries depict different connotations. For example, a traditional costume in Central Asia is still very much alive and worn in everyday life, partly because it is the most practical wear for this region and partly because it fosters national pride (Traditional Costume of The Republics of Central Asia 2014). An African dashiki costume symbolizes Black Nationalism (Valerie 2005i) whereas Chileans' national costume fosters festivity and convenience (Traditional

Fig. 24 Tracht (*Source* https://commons.w ikimedia.org/wiki/ file:Stadtgruendungsfest_ munich_2013_Paar_in_Trach t_beim_TAnz.jpg)

Costume of Chile 2015). The Tracht (Fig. 24), considered as a pure traditional costume in Germany, played a vital role in the resistance movements in the country, as it was an attempt to establish itself as an icon of national identity or **identical emblem of Germanic people** (see Footnote 9).

In Ghana, the type of fabric used, the motifs, and colors used with ethnic fabrics help in expressing the views, prestige, and status of the wearer (Bernard et al. 2013). Korean traditional dress stands for their traditional values, such as philosophy, religious attitude, and family relationships (DeLong and Key Sook 2009). Indonesia's national costume, the kebaya (Fig. 25) symbolizes national pride and feminist values.[12] Batik is recognized as one of the important identities of Indonesian culture. UNESCO designated the Indonesian batik as a "Masterpiece of Oral and Intangible Heritage of Humanity" on October 2, 2009 (see Footnote 3).

In Vietnam, symbolically, the Ao dai (Fig. 26) invokes nostalgia and timelessness associated with a gendered image of the homeland for which many Vietnamese people throughout the diaspora yearn for (see Footnote 4). Similarly, the Mexican poncho represents connotations of power among the Mapuche population; the stepped-diamond motif was considered to be a sign of authority and

---

[12]http://www.expat.or.id/info/kebayatraditionaldress.html. Accessed 01 December 2015.

Fig. 25 Indonesian kebaya
(*Source* https://commons.wik
imedia.org/wiki/file:Baju_ke
baya.jpg)

Fig. 26 Ao dai (*Source*
https://commons.wikimed
ia.org/wiki/file: Aodai_in_
Purple.jpg)

Fig. 27 Japan's modern day furisodes (*Source* https:// commons.wikimedia.org /wiki/File:TokioShibuya-Kimono.jpg)

was often only worn by older men, leaders, and the heads of the paternal lineage in families (Valerie 2005b).[13]

2.1.7 Effect of Acculturation

Ethnic dress has served as cultural symbols of particular social groups, but what is important is, have these nurtured values sustained the imperialistic influences of capitalist Western societies because today every society faces acculturation. Certainly, this acculturation has affected ethnic dress patterns in several ways.

In Japan the modern day furisodes (Fig. 27) feature contemporary versions of traditional floral patterns (see Footnote 2). In Indonesia the intricate batik printed fabrics have been converted into contemporary shirts, and the pareu comes in all vivid hues and patterns (see Footnotes 3 and 5). In India the traditional kurtas and kurtis are paired along with chinos and denim alike. Today lawn fabric is made from cotton and in a variety of color combinations and it is often embellished with modern motifs such as splashes of color, geometric prints, swirls, and similar.[14]

---

[13]http://www.donquijote.org/culture/mexico/fashion/the-poncho. Accessed 01 December 2015.

[14]http://www.utsavpedia.com/attires/lawn. Accessed 30 November 2015

Fig. 28 Mexican poncho with kasuti embroidery (*Source* https://flickrhivemind.net/blac kmagic.cgi?id=8640442514)

Fig. 29 Contemporary flamenco costume (*Source* https://zumodekiwi.files.wor dpress.com/2013/01/ verde-y-negro)

The Mexican ponchos (Fig. 28) are currently used as waterproof cloaks and as fashion items during autumn and winter in western countries (see Footnote 13). Over time, the Flamenco costume (Fig. 29) has become richer in color and has adopted adornments and complements such as lacing, embroidered ribbons, flowers, costume jewelry, and hand fans.[15] The African dashiki has embraced several

---

[15]http://www.donquijote.org/culture/spain/flamenco/spanish-flamenco-dress. Accessed 01 December 2015.

informal and formal costumes, namely the black tie, white tie, and morning dress.[16]

Starting in 1910, led by Paul Poiret and Jeanne Paquin, the ideas of Eastern dress have come to be a part of Western dress, including saris and dhotis from India, kimonos from Japan, caftans and djellabahs from North Africa, and cheong-sams from China (see Footnote 11). However, it is not always the same; the reverse process with Western clothes entering the households of Asia and Africa is even greater. The media of magazines, television, and cinema have allowed the people of Asian origin to be exposed to a whole host of ideas, images, and prac-tices from across the globe. Rather than the end result being Westernization, these various ideas and practices interact with the cultural influences already present in people's lives (such as their Indian traditions) to produce cultural diversity and heterogeneity (Sertanya 2009).

2.1.8 Sustainability of Coloring Techniques

The coloring process, more technically known as dyeing, is the most cumber-some and also influences the design outcome. The application of color to a textile substrate has seen a lot of phenomenal innovations and has been provided with a vastly expanded color palette beyond comprehension. Neon colors would prob-ably never have been imagined or thought of in the early 1900s. In traditional cos-tumes, good old methods with the help of customized newly fabricated machines still hold sway. The traditional craftsmen still follow the tedious and cumbersome processes to produce characteristic aesthetic effects and affordability, incurring huge capital investments, leading to highly customized designs.

In traditional Mexican clothing, indigo (blue), brazil wood and cochineal (red), palo de tinta (black), cinnabar (red-brown), and purpura patula (lavender) may have been in use (Odland 2006). However, in 1856 the invention of chemi-cal dyes in Europe expanded the color palette throughout the world. These dyes were quickly adopted and used along with some of the natural dyes. By the early 2000s, natural dyes were reintroduced to many Mexican and Guatemalan weavers and embroiderers (Odland 2006). Rainbow coloring is a predictable and enjoyable aspect of twenty-first-century clothing.

Traditionally, kimonos are hand painted on a rice paste resist base. Even vast areas of background are dyed in this manner, i.e., by hand painting. The most characteristic features of kimono ornamentation made in this way produce subtle color gradations and narrow, flowing light lines that outline the motifs. Japanese ikat are made by selectively binding and dyeing parts of the warp or weft threads, or even both, before the fabric is woven.[17] It is an arduous and exacting process

---

[16]http://www.thewrendesign.com/the-story-behind-african-wax-print-cloth.   Accessed   30 November 2015

[17]http://www.howdesign.com/design-creativity/how-to/kimonomaking. Accesed 28 November 2015.

Fig. 30 Batik print (*Source*
The process of Batik:
klowong, isen and harmonic
ornamentations (Doellah
2002)

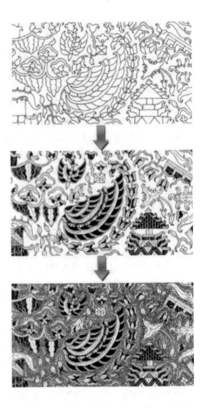

where the threads are stretched on a frame, selected design areas are bound, then the hanks of bound threads are immersed in the dye pots using natural dyes. The Nishijin style of weaving uses yarn dyeing, in which yarns of various colors are woven to make patterns. This technique is both time consuming and labor intensive compared to other techniques, but it is indispensable for creating the elaborate and gorgeous designs required for kimono fabric.

Indonesian batik printing (Fig. 30) also uses a wax resist printing technique. After the initial wax has been applied, the waxed fabric is immersed in the dye bath of the first color. The amount of time it is left in the bath determines the hue of the color; darker colors requiring longer periods or numerous immersions. The fabric is then put into a cold water bath to harden the wax. The number of colors in the batik represents how many times it was immersed in the dye bath and how many times wax had to be applied and removed (see Footnote 12). A multicolored batik represents a lot more work that a single or two-color piece. African dashiki prints also use the same wax resist dye process but the motifs are much more elaborate.

In Peru **Yanali**, almost all colors used in yarn dyeing are natural dyes obtained from a variety of natural sources available locally. **Cochineal, a scale insect of Central and North America** is the most commonly used substance for the production of red dye (Antúnez de Mayolo 1989). Ch'illca, a green leafy plant with

Fig. 31 Bandhini work in India (*Source* https://commons.wikimedia.org/wiki/File:Bandhani_print_open.jpg)

white flowers, serves as the biggest source of green. The plant Mutuy is another source for producing a dark green color, and this extremely dark green is usually over-dyed with yellow, sourced from plants such as Ñuñunqa, to produce a much lighter, brighter shade of green (Antúnez de Mayolo 1989).[18] For producing purple, cochineal substance and a fixative, such as copper or iron oxide, are mixed. Colors such as blue and gray are produced by combining substance **Tara**, a bean-like pod and blue collpa (usually a local form of copper sulfate) (see Footnote 18) (Antúnez de Mayolo 1989). Indigo is still used for blue. The Pareu is traditionally hand painted in Polynesia with fabric paint. The themes usually used are flora (tropical flowers, breadfruit trees called "maiore"), sea life, and the islands—with quite bright colors (see Footnote 5) (Aloha Shirt History 2014; Philip 2005b).

Bandhini work (Fig. 31) is one of the most elegant ethnic works in India. This process begins with tying the knots and then the fabric is thoroughly washed to remove the imprint. The cloth is then dipped in naphthol for 5 min and dyed in yellow or another light color for 2 min. Next it is rinsed, squeezed, dried. and then tied again and dipped in a darker color. It is kept there for 3–4 h (without opening the knots) to allow the color to soak in. During this process the small area beneath the thread resists the dye, leaving an undyed spot. This is usually carried out in several stages starting with a light color such as yellow, then after tying some more knots a darker color is used and so on. After the last dyeing process has been completed the fabric is washed and, if necessary, starched. After the fabric is dried,

---

[18]http://www.incas.org/category/weavings/making-and-dyeing-yarn. Accessed 29 November 2015

Table 1 Typical effluent characteristics

Determinant	Woven fabric finishing	Knit fabric finishing	Stock and yarn dyeing and finishing
BOD[a] (mg/l)	550–650	250–350	200–250
Suspended solids (mg/l)	185–300	300	50–75
COD[b] (mg/l)	850–1200	850–1000	524–800
Sulfide (mg/l)	3	0.2	0–0.9
Color (ADMI[c] units)	325	400	600
pH	7–11	6–9	7–12

[a]Biological oxygen demand
[b]Chemical oxygen demand
[c]American dye manufacturers Institute
Source GG62 guide, Good guide practice to water and chemical usage in the dyeing industry, First printed 1997

its folds are pulled apart in a particular way, releasing the knots and revealing their pattern. The result is usually a deep colored cloth with spots of various colors forming a pattern (History of Bandhani or Indian Tie and Dye Technique 2012; Sunetra 2011).

Typically, the ethnic color patterns come in rich and vivid hues with each color concept having its own story to tell. Minimalistic features, monotonous schemes, or color blocking schemes literally never existed in ethnic dress. Almost all the coloring techniques used in the preparation of ethnic dresses are either printing or yarn dyeing techniques. In yarn dyed goods of ethnic costumes, natural dyes are still used by a few craftsmen although conventional dyes have replaced them by and large. The effluents from conventional yarn dyeing and finishing processes are comparatively less than from woven fabric finishing and knit fabric finishing. Similarly, using artisanal crafts in converting yarns into fabric from locally available materials drastically cuts the carbon emissions generated through raw material transport and industrial machines (Alicia et al. 2013). Table 1 details the typical effluent characteristics of the dyeing and finishing industry.

2.1.9 Artisanality of Ethnic Dress

Hawaiian aloha print (Fig. 32) motifs are geometrical, natural, abstract, or adapted to style. Nevertheless, they all depict a tropical theme with different permutations of hues and color contrast themes. In the early days the original aloha shirt designs were comprised of floral patterns of the local countryside (Aloha Shirt History 2014; Paula 2011). The regular visits of Western tourists witnessed the adapted styles of desired color patterns (Aloha Shirt History 2014). Today one can see all the seven contrasts developed by Joannes Itten, such is the artistry levels we have reached in printing. Mexican blouses, a testimony to the quintessential Bohemian

Fig. 32 Hawaiian aloha print (*Source* https://commons.wikimedia.org/wiki/File:Vintage_aloha_shirt.JPG)

Fig. 33 Dashiki print design (*Source* https://commons.wikimedia.org/wiki/File:Dashiki_and_kufi.jpg)

look, comprise a vast range of artisanal elements with each variant of blouse featuring either single or a combination of beautifully decorated laces, beads, colorful patterns, and embroidery (Traditional Mexican costume 2014).

The dashiki print design (Fig. 33) comes in medium and oversized, clearly defined patterns placed against a large contrasting color background (space). It promotes vibrant, playful schemes that can be engineered as desired. The big

Fig. 34 Ethnic Indonesian batik prints (*Source* https://commons.wikimedia.org/wiki/
File:Printing_wax-resin_resist_for_Batik_with_a_Tjap,_Yogyakarta,_1996.jpg)

patterns are marked by the design elements, lines and shapes in an architectural
style connoting values of individualistic expression (see Footnote 7).

Ethnic Indonesian batik prints (Fig. 34) and Indian and Chinese fabric print patterns constitute a map of repetitive intricate icons or multiple intricate icons telling
a collective story. These intricate icons are block printed and can be enriched by
overlapping/superimposing fine hand painting techniques. In some cases they are
also enriched by adding embellishments such as beads and sequins. In a few other
styles, such as ikat weaves, differently colored warp and weft are juxtaposed to
each other in a measured way to produce defined pattern stories. Similarly, needlecraft is also used to enrich the fabric appearance, e.g., zardosi embroidery and
chikankari embroidery of India. Every craft skill under the sun has been used to
enrich fabric or garment appearance. At the same time, concentration on shaping
the pattern pieces to create a durable garment expressing either rigid structural
lines or smooth flowing drape lines along with decoration techniques was seldom
considered in these ethnic dress styles. Wrapping, draping partially tailored panels, tying, and knotting was the norm. All the artisanality and craft of these ethnic dresses ensured flamboyant expression of icons in minute detail to convey a
culturally valued outlook. Unlike this scenario, France in particular nurtured traditional craftsmen in pattern making, tailoring, and needlecraft decoration techniques which today stands as the epitome of haute couture dress.

However, a few costumes in the world such as Ladakhi goncha and Greenland's
national costume rival the craft and finesse of haute couture even today and match
the sophisticated look of contemporary holiday suiting. Greenland's national costume (Fig. 35) in terms of artistic quality and exuberance is replete with all the
features of a classical suit, the national costume being an ensemble of bead tops,

Fig. 35 Greenland's national costume (*Source* https://commons.wikimedia.org/wiki/File:Upernavik_first_day_in_class_2007-08-14_2.jpg)

thigh length knickers, lace stockings, and boots made of sealskin.[19] Similarly, ladaki goncha (Fig. 36) is an ensemble comprising thick woolen cloth with a colorful cummerbund tied at the waist, loose pajamas, a top hat, and long felt boots (Valerie 2005m). Another beautiful feature of this ladaki goncha is the head-dress perak studded with around 100–400 semi precious stones and ornamental jewelry.

2.1.10 Innovations Using Ethnic Craft Skills

Iconic fashion designers either used ethnic artisanality, tailoring skills, advanced technologies, or combination of more than one skill as tools in turning out exclusive designer garments. The scope of adaptability is also high for ethnic artisanality skills as they can be easily manipulated to fit current requirements.

Designer Manish Arora uses psychedelic color patterns and elaborate embellishments of his native India in all of the designs signifying his design philosophy. These distinctive Indian design aesthetics are deployed on Western silhouettes, thereby transforming into vibrant eclectic garments (Alicia et al. 2013). Designer Alabama Chanin employs artisans for converting high grade organic and recycled materials into fashion garments. These fashion garments also bear the artisans signature on them (Alicia et al. 2013). Alber Elbaz is considered a master in creating fashion garments emphasizing flow and volume fashion principles. Alber Elbaz utilizes old world tailoring materials and dressmaking skills to form seamless

---

[19]http://www.greenland.com/en/about-greenland/culture-spirit/traditional-dress.   Accessed   01 December 2015.

Fig. 36 Ladaki goncha
(*Source* https://commo
ns.wikimedia.org/wiki/
File:Ladakh1981-326.jpg)

draping and pleating with special attention to fine details, pristine finishing, and
eye-catching jewelry coordinators (Alicia et al. 2013).

Carla Fernandez is widely known for making fabrics on a backstrap loom.
Designs are basically made using geometric pattern pieces, primarily squares and
rectangles similar to traditional Mexican garments, such as tunics and shawls with
the objective of minimizing the need for cutting the fabric; folds, darts, tucks, and
pleating are deployed to tailor the garment (Alicia et al. 2013). This approach is
used in the production of the entire range of softly draped silhouettes to boldly
architectural ones. Ralf Rucci is inspired by Asian culture and contemporary art
alike, as reflected in his custom produced fabrics (Alicia et al. 2013). Ralf Rucci's
designs tell tales of exotic bohemianism. Designer Van Noten employs unusual
textiles and places overlapping prints on already patterned fabrics (Alicia et al.
2013). The other notable elements of his collection include opulent embellishments of beading or embroidery and crisp cotton against the cool sheen of snake-
skin. The artisanality resembles a Namibian Heroro woman's costume patterns.

Designer Castellano's handbags (Fig. 37) are made by the Wayuu people, who
live in the desert state of La Guajira, Colombia and the Arhuaca people, who live
in the mountainous areas in the north (Chere 2015). These people are traditional
craftsmen trained in sewing and weaving for centuries. Designer Fiona Paxton's

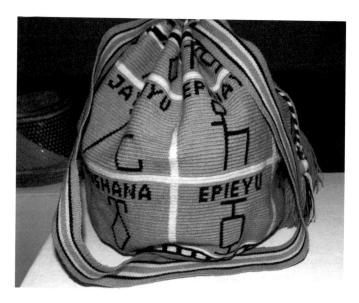

Fig. 37 Castellano's handbag (*Source* https://commons.wikimedia.org/wiki/File:Susu_con_tejido_de_shi'iruku.jpg)

Table 2 Artisanal fashion and outcomes

Designer	Process	Outcomes
Holistic role	In-house and local	Product satisfaction
Integrated skills: design and craftsmanship	Small-scale, individually crafted, and unique	Good fit, high quality, and durability
Control over process and products	Traditional and handcraft methods	Lower environment impact
Frequent communication with team members	Teamwork	Ethical production

obsession with art and ancient craftwork has helped her to redesign the jewelry of the modern traveler using tribal motifs and traditional patterns.[20]

Organization ABURY foundation combines traditional, old world crafts with avant garde designs to create a new luxury style. ABURY brings together exciting designers with traditional artisans from remote and inspiring cultures.[21] ABURY's process involves the practice of training designers in a craft culture for roughly 3 months to learn and share ideas from one another and to create a collection together, bringing the best of heritage knowledge and wisdom to high design and sustainable solutions. Table 2 represents artisan fashion, processes, and outcomes. Ethnic styles can be best developed with a combination of good designer roles,

---

[20]http://www.fionapaxton.com. Accessed 01 December 2015

[21]http://abury.net. Accessed 15 November 2015

best processes that aim at product satisfaction with good fit, high quality, durability, lower environmental impact, and ethical production.

2.2 Scope of Ethnic Costumes for the Next Generation

2.2.1 Fashion Systems Model

After the First World War, with the onset of technological developments and industrial capitalism, both men and women started working in factories. Homogenous styles, muted solid colors, and minimalistic patterns with color blocking schemes were mass produced on a large scale. The wide range of fabric printed designs inspired by different art movements and paintings alike fell out of favor. The story goes that merchants traveled to France to look for print patterns and paintings that could be translated into fabric print designs in their mechanized machines. A new fashion system was coded to exemplify the details of appearance and attributed symbolic meanings that reflected a person's character or social standing (Valerie 2005j).

Post-1950s, fashion has witnessed the rise of subcultures that broke the totalitarian fashion concepts and paid attention to small close group values (Valerie 2005j). Slowly fashion became decentralized from the perspective of structural functionalism to symbolic interactionism. This phenomenon is accepted as postmodernism in which expression of self takes precedence over conforming to a dressing style (Valerie 2005j).

2.2.2 Generation Z

Generation X style characteristics can be inferred from a few signature style looks: metro sexual or grunge characterized by untidy, shabby looking hair-styles, which are trendy. By far, for many youngsters the prevailing style is jeans, cool sneakers, and messy hair for men and jeans, cool sneakers, and neat hair for women.

However, Generation Z youngsters wish to express their own style in an anonymous manner.[22] Generation Z is also nicknamed Digital natives, just because they can simultaneously create a document, edit it, post a photo on instagram, and talk on the phone. Other elements of Generation Z include, rave look, pop art style skirts printed with a giant coca cola logo, tie-dyed maxi dresses, and rainbow chokers bearing a resemblance to Generation X.[23] In addition to this, the 2015 spring beauty trend report of Teen Vogue indicated that with the vintage Patty Smith style, shag haircuts, and ripped fishnets era, and scarlet lipstick as trends,

---

[22]http://www.nytimes.com/2015/19/20/fashion. Accessed 29 November 2015

[23]http://www.Teenvogue.com/gallery/spring-summer-2015-beauty-trend-report. Accessed 29 November 2015.

the youth style bloggers are latching on to anonymous styles pertaining to smaller sub-groups and sub-cultures, an example being the self-defined laundry-day look that comprises oversize sweaters, baseball caps, and jogging pants (see Footnote 22). This clearly reveals that there is no single acclaimed and sought after signature look mimicking pop stars and cultural stars. The current generation's obsession with "Selfies" can only increase the Generation Z's presence in the ever increasing social networking sites.

2.2.3 Scope of Ethnic Dress

Anonymous style and sub-cultures are the order of the day; the fashion personality is rich and diverse. Rich in the sense of pluralistic visual concepts such as a combination of vintage and modern art styles on the fashion scenes. A classic example is the fashion style of Generation Z icon Miley Cyrus, who has left the fashion forecasters bewildered in getting the terms right for Generation Z. Meanwhile, the ethnic palette is replete with geometric shapes, plants, animals, and human images woven in a representational, stylized, or abstract fashion. These iconic designs are symbolic and are rooted in the collective consciousness of the local people. In an era of acculturation and emboldened selves, neither distinguished citizens nor laymen of a country would blindly imitate a foreign ensemble outright as it would only result in cognitive dissonance. The fashion identity is seen as a projection of one's self, and hence the search for a local element in fashion clothing and the relative associations drawn in the cultural perspective is important. The ethnic concept avails itself a chance to co-exist with modern fashion concepts.

3 Conclusion

Ethnic styles are stand-alone costumes that display the cultural values and clothing style of a particular race or community. Ethnic styles have evolved with geographical regions: the tunic and salwar are popular in central Asia; East Asia is the place for cheongsam and mao suit (China), kimono (Japan), hanbok (Korea), sarong (Indonesia), and Ao dai (Vietnam); salwar kameez, ghagra/lehenga choli, pajamas, pirans, mekla chaddar, sarees, dhoties, and gho are popular in South East Asia; the Chilean chamanto, Peruvian poncho, and Inca dress are popular in Latin and Central America; sarape, charro suit, sombrero, guayabero, baja jacket, and poncho are popular in North America; and folk dress (Portugal), flamenco dress (Spain), and tracht (Germany) dominate Europe.

Artisanality has contributed to the evolution of ethnic styles which can be evinced from the way Ladakhi woman's dress and Greenland's Inuit's dress have created attention and are considered on a par with the work of fashion innovators such as Ralph Rucci, Dries Van Noten, Nicolas Ghesquiere, etc. The role of designers in the evolution of ethnic styles is paramount and designers such as

Fiona have turned tribal patterns into statement fashion jewelry with a versatile look, whereas designer Carla Fernandez has turned the traditional Mexican tunic and shawls into contemporary fashion clothes using ancient techniques and provided a means for social empowerment of the craftsmen. Organizations such as ABURY provide a platform for traditional craftsman to exhibit their stuff and protect their interests.

Innovations combined with sustainable manufacturing practices such as the use of eco-friendly raw materials such as natural fibers, natural dyes, and recycled materials and the use of resist bases, hand painting, itkat, batik, bandhi, natural dyes and effluent treatments have added another level to ethnic styles. Generational changes are expected to play a vital role in the evolution of ethnic styles. Typically, ethnic styles are fixed costumes, which is quite a contrast to the cyclic patterns of fashion phenomenon. Ethnic patterns and styles are a source of strength, pride, and identity to fashion-seeking youngsters. However, the characteristics sought by an individual drive them to push forward his or her indigenous value (Wicklund and Gollwitzer 1982), which simultaneously determines the symbolic interactions while stating fashion clothing preferences. The future of ethnic dress/style depends on peoples' interactional experience combined with their personalities and individual behavior.

References

Alicia K, Emily BS, Jay C (2013) Fashion design visual referenced, 1st ed. Rockport publishers, Beverly, pp 202, 357, 364, 384, 398

Aloha Shirt History (2014) http://www.alohashirtsofhawaii.com/alohashirthistory. Accessed 01 Dec 2015

Arti S (2014), Indian fashion: rradition, innovation, style, 1st edition, Bloomsbury Publishing, p 44

Bernard ED, Robert A, Richard G (2013) In: Arts and design studies, vol 13. Available via DIALOG. http://www.iiste.org. ISSN 2225-059X (Online). Accessed 26 Nov 2015

Carmen L (2002) CZECH/SLOVAK PAGEANT (folk wear) GUIDELINES. Available via DIALOG. http://www.acscc.org/pdf/KrojGuidelines_020507.pdf. Accessed 28 Nov 2015

Central Asian Women's Adornment and Clothing (2014) http://www.wdl.org/en. Accessed 28 Nov 2015

Chere DB (2015) Folk heroes: artisanal clothing with ethics. http://eluxemagazine.com/fashion/folk-heroes-artisanal-clothing-with-ethics. Accessed 28 Nov 2015

Eicher JB (1995) Introduction: dress as expression of ethnic identity. Oxford; Washington, DC, pp 1–5

Eicher J, Sumberg B (1992) World fashion, ethnic, and national dress. Dress and ethnicity. Oxford, Washington, DC, pp 297–367

History of Bandhani or Indian Tie and Dye Technique (2012) http://theindiacrafthouse.blogspot. in. Accessed 25 Nov 2015

http://www.marlamallett.com/k_design.htm. Accessed 27 November 2015

http://www.turismoinkaiko.net/peru-travel/peru-culture-traditions/traditional-dress-in-peru.html. Accessed 30 November 2015

Indian Traditional Costume and Makeup (2010) http://hinduonline.co/HinduCulture/IndianTradit ionalCostume.html. Accessed 25 Nov 2015

Antúnez de Mayolo KK (1989) Peruvian natural dye plants. Econ Botany 43(2): 181–191. Available via DIALOG. www.jstor.org/stable/4255151. ISSN 00130001

DeLong MR, Key Sook G (2009) Korean dress and adornment. http://fashion-history.lovetoknow. com/clothing-around-world/korean-dress-adornment. Accessed 01 Dec 2015

Odland JC (2006) Fashioning tradition: Maya Huipiles in the field museum collections: Fieldiana, Anthropology, New Series, No. 38. Biodiversity Heritage Library Collection

Li Y (2009) A study on the application of elements of ethnic dress in modern fashion design. Can Soc Sci 5(2):70–73

Paula R (2011) Fine print. Kahala 6(1). Available via dialog. www.kahalaresort.com/i/downloads/ Magazine__2011.pdf

Peruvian National Costume (2015) Woven clothing from Alpaca wool, unique patterns and bright colors. http://nationalclothing.org/27-nationalclothing/america/peru. Accessed 28 Nov 2015

Philip S (2005a) A history of fashion and costume—the Nineteenth Century. Bailey Publishing Associates Ltd., London, pp 13–19, 29–30

Philip S (2005b) A history of fashion and costume—the Nineteenth Century. Bailey Publishing Associates Ltd., London, p 5

Philip S (2005c) A history of fashion and costume—the Nineteenth Century. Bailey Publishing Associates Ltd., London, p 28

Sertanya R (2009) Styling the self: fashion as an expression of cultural identity in a global world. Communication and Media Studies Department University of KwaZulu-Natal, South Africa

Sunetra D (2011) Evaluation of bandhani motifs and processes of rajasthan: from past to contemporary. In the Faculty of Social Sciences: The IIS University Jaipur, pp 2–18

Sunita P (2001) Ethinc fashion obscures cultural identity, The Yale Herald, 2 Feb

Traditional Costume of The Republics of Central Asia (2014) http://local-moda.blogspot. in/2014/06/traditional-costume-of-republics-of.html. Accessed 26 Nov 2015

Traditional Costume of Chile (2015) Flowered dresses, chamantos and chupalla hats. http://national clothing.org/30-nationalclothing/america/chile. Accessed 28 Nov 2015

Traditional Mexican Costume (2014) Typical pieces of clothing in Mexico http://nationalclothing. org/21-nationalclothing/america/mexico/18-traditional-mexico. Accessed 26 Nov 2015

Valerie S (eds) (2005a) Encyclopedia of clothing and fashion, vol 1. Thomson Gale, New York, p 22

Valerie S (eds) (2005b) Encyclopedia of clothing and fashion, vol 1. Thomson Gale, New York, pp 37–48

Valerie S (eds) (2005c) Encyclopedia of clothing and fashion, vol 1. Thomson Gale, New York, pp 79–80

Valerie S (eds) (2005d) Encyclopedia of clothing and fashion, vol 1. Thomson Gale, New York, p 83

Valerie S (eds) (2005e) Encyclopedia of clothing and fashion, vol 1. Thomson Gale, New York, p 84

Valerie S (eds) (2005f) Encyclopedia of clothing and fashion, vol 1. Thomson Gale, New York, p 90

Valerie S (eds) (2005g) Encyclopedia of clothing and fashion, vol 1. Thomson Gale, New York, p 94

Valerie S (eds) (2005h) Encyclopedia of clothing and fashion, vol 1. Thomson Gale, New York, p 95

Valerie S (eds) (2005i) Encyclopedia of clothing and fashion, vol 1. Thomson Gale, New York, p 347

Valerie S (eds) (2005j) Encyclopedia of clothing and fashion, vol 2. Thomson Gale, New York, pp 21–25

Valerie S (eds) (2005k) Encyclopedia of clothing and fashion, vol 2. Thomson Gale, New York, p 95

Valerie S (eds) (2005l) Encyclopedia of clothing and fashion, vol 2. Thomson Gale, New York, pp 100–101

Valerie S (eds) (2005m) Encyclopedia of clothing and fashion, vol 2. Thomson Gale, New York, pp 235–240

Valerie S (eds) (2005n) Encyclopedia of clothing and fashion, vol 2. Thomson Gale, New York, pp 246–247

Valerie S (eds) (2005o) Encyclopedia of clothing and fashion, vol 3. Thomson Gale, New York, p 4

Wicklund RA, Gollwitzer P (1982) Symbolic self completion. Routledge, London, p 31

Continued Change in *Geringsing* Weaving in Tenganan, Bali

L. Kaye Crippen

Abstract Traditional ethnic textiles make the world a more interesting place and a respite from the uniformity of the mass produced textiles of fast fashion. The richness of watching the continuation or revival of traditional textile production in person and the resulting use of apparel and other textile products used in daily, ritual, or ceremonial aspects of life, either sacred or profane, remind us of the beauty and importance of traditional ethnic textiles. As more tourists experience traditional textile cultures first hand, it is not surprising that there is a link between tourism and helping with the continuation of the production of ethnic textiles. The internet sale of textiles has also had a positive effect on the demand for traditional woven textiles. The current popularity of *ikat* patterned textiles or imitation ones, i.e., printed, is discussed. This chapter documents continued changes in weaving in the village of Tenganan Pegeringsingan in Bali, Indonesia from 1985 to 1999 and 2014. The village has a century's old tradition of producing difficult to make *geringsing* double *ikat* textiles which requires tie-dyeing of both warp and weft yarns in both the warp and weft directions to create a pattern. Reasons for the decline and subsequent partial revival of weaving are explored.

Keywords Bali · Double *ikat* · Ethnic textiles · *Geringsing* · *Ikat* · Resist dyeing · Tenganan pegeringsingan · Tie-dying · Traditional textiles

L. Kaye Crippen (✉)
University of Arkansas at Pine Bluff, Human Sciences,
1200 University Drive, Pine Bluff, AR 71601, USA
e-mail: crippenk@uapb.edu

© Springer Science+Business Media Singapore 2016
M.A. Gardetti and S.S. Muthu (eds.), *Ethnic Fashion*, Environmental Footprints
and Eco-design of Products and Processes, DOI 10.1007/978-981-10-0765-1_7

1 Introduction

1.1 Traditional Textiles Using Resist Dye Techniques

Traditional or ethnic textiles use a wide range of textile production and adornment techniques which were used before the industrial revolution. Some traditional textiles have been updated to incorporate new technologies. Interest in traditional or ethnic textiles has increased rapidly, starting in the 1960s with the hippie era when tie-dyed textiles and other ethnic apparel, i.e., African *dashikis* were *de rigueur*.

The cult of tie-dye has simmered since the 1960s but the trend has strengthened recently for apparel, accessories, and interior textiles. It also brought with it a fascination for *ikats*—both authentic and imitation—and other traditional textile techniques including felting, which has reappeared on the fashion scene recently. Both tie-dye, an American term, and *ikat* are classified as resist textiles because they use some media to make the yarn or textile resist dye penetration. The major types of resist textiles are *ikat* (pronounced 'ee-KAHT), batik, and tie-dyed textiles, the operation being done on the textile or textile product. All of these have different names in different countries.

The *ikat* technique shown in Fig. 1 is classified as a warp *ikat* because only the warp yarns were tied before dyeing which resists dye from penetrating in areas tied off or bound in the warp direction.

The resulting light spaces in Fig. 1 were undyed as a result of having something tied around the area to resist the blue indigo dye in which they were dipped; the resulting dark pattern becomes evident once the textile is woven, in this case on a back strap loom.

Other variations of the *ikat* technique are discussed below. *Ikat* textiles are known for their indeterminate or hazy lines. True *ikats* achieve this characteristic look by tying the yarns before dying; however, printing on either the warp yarns or the finished cloth can result in an *ikat* look that novices sometimes refer to as *ikats*.

When a woven textile or a textile product, i.e., scarf, blouse, skirt, or drapery textiles, are dyed as a textile or product using resist techniques, it is called tie-dying. In Indonesia this is called *plangi* or *tritik*, in India it is known as *bhandari* or *bandhani* as shown in Fig. 2, and in Japan it is *shibori*.

Batik textiles use wax, rice paste, or other media to draw, paint, or print on the textile to resist the dye penetration. A batik may have two or more colors. In Indonesia, various types of waxes are applied using a hand held pen type instrument, the *canting* (pronounced chant-ing), which has a reservoir for a small amount of hot wax and a hole for dispensing the wax. This allows the batik maker to draw on the textile with hot wax which resists dying. The resulting fabrics are called batik *tulis* (pronounced too-lease) which are the most prized of the batiks. Various types of instruments are used in different countries, e.g., China.

Batik *cap* (pronounced chop) in Indonesia refers to both the copper instrument, a *cap*, as well as the resulting textile. The *cap*, which is dipped into wax and then

Fig. 1 Fuzzy lines from figure on new Sumbanese warp *ikat* from author's collection purchased at TOL (2015). Photographer W. Hehemann

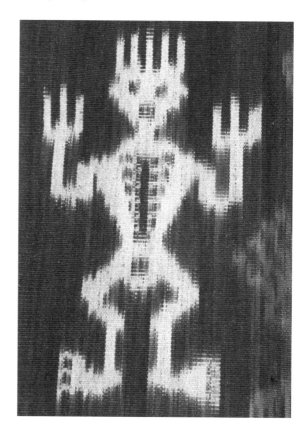

stamped onto the textile, being careful to carefully align or register the wax, makes the resist pattern. This greatly speeds production time and the product is not considered as valuable as batik *tulis*. Most of the batik *tulis* is made by women in Indonesia; *cap* production is done by males.

2 Imaging *Ikat*

Today, there is heightened interest in authentic *ikat* such as those offered on websites, e.g., Cloth Roads https://www.clothroads.com located in the U.S. in the state of Colorado; their web-site promotes authentic *ikats* as described by Crippen and Mulready (2014). Their blog on their web-site helps to educate consumers on where and how the products are made and where they are sourced.

Threads of Life (TOL) in Ubud, Bali in Indonesia, which has a retail store in Ubud, Bali, has recently started their internet business at http://threadsoflife.com/shop/. A detail from a Sumbanese *selandang*, an Indonesian shoulder scarf, is a reproduction from TOL (see Fig. 1). This is characteristic of the indeterminate

Fig. 2 Tie-dye *bandini*
from author's collection
purchased from Judy Frater
Somaiya Kala Vidya 2015.
Photographer W. Hehemann

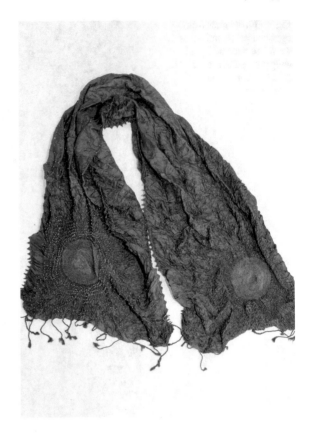

nature of the fuzzy lines characteristic of *ikats*. This textile was dyed blue using indigo where the dark color is located and is undyed where the yarns were tied to resist dye penetration. It is a warp *ikat*.

It was a large Sumbanese *hinggi*, a wide double paneled textile worn by men, which the author first encountered in the early 1970s on a visit to the Great Western Flea Market in Los Angeles with friends in search of American coverlets, traditional hand-woven linen or cotton, and wool bed coverings. As friends continued onward, the author stopped in her tracks, mesmerized by this large textile. The seller asked $25.00 for it. Sold! No bargaining here. Luckily it had a small piece of twill tape labeled with ink stapled to the selvage which was removed for conservation purposes that said Sumba. Where's that?

The University of California at Los Angeles (UCLA) Art Library had one of the few books on *ikat* textiles at the time by M. J. Adams, an anthropologist and researcher at the Peabody Museum of Archaeology and Ethnology (http://www.peabody.harvard.edu/), entitled *System and Meaning in East Sumba Textile Design* (1969). At this time the Los Angeles County Museum of Art (LACMA) Costume and Textile Department had only seven books with entries on *ikat* in their index. Bühler wrote several articles on natural dyeing and *ikats* including

"*Patola Influences in Southeast Asia*" (1959). Bühler et al. (1975/1976) wrote on *patola* and *geringsing* textiles. Ramseyer has also written extensively on Tenganan Pegeringsingan (Ramseyer 1984; Ramseyer and Hinz 1977). Ramseyer's most recent book *The Theatre of the Universe: Ritual and art in Tenganan Pegeringsingan* (Ramseyer and Berger 2009) offers the most comprehensive view of this ancient village of any book. Gittinger's *Splendid Symbols* (Gittinger 1985) offered numerous insights. Today the proliferation of books on *ikat* textiles continues (Barnes and Kahlenberg 2010; Bebali 2011; Hamilton 1994, 2012, 2014; Maxwell 2013) and there are many others too numerous to mention who have greatly contributed to our understanding of these traditional *ikat* textiles.

The Textile Museum in Washington, D.C. announced in a press release (Anon 2015) that Guido Goldman had donated 76 Asian *ikat* panels and textiles. This collection, documented in the book *Ikats: Silks of Central Asia* (Gibbon and Hale 1997) gave extensive insight into these *ikats*.

In Margalin, Uzbekistan in 2012, a woman sells silk velvet *ikats* in Fig. 3 and a woman in the same city weaves warp *ikat* on her loom in Fig. 4. There are also men weavers in this geographic area; however, men do all the tying of the yarns.

Fig. 3 Woman and daughter sell silk *ikats* in Margilan, Uzbekistan. Photographer M. Connors

Fig. 4 A woman weaves
warp *ikat* textile on her
loom in Margilan, Uzbekistan

From this same city is a scarf from the author's personal collection purchased from Cloth Roads; Cloth Roads' blog has information on *ikats* from the workshop that wove this piece (Ludington 2011) (Fig. 5).

An apparel design professor who saw this textile during the photo shoot thought that this textile had been digitized and copied; however it was a new authentic traditional textile motif from this region. This example illustrates how difficult it is for consumers and design/textile professionals to understand traditional textiles. Education of all of these is important so they can begin to understand what is authentic if that is important to them and what is not, so they can make informed decisions. It also helps the marketing function if named patterns inspired by certain techniques, regions, or cultures are used, together with more authentic names and traditional textile terms such as *ikat*.

Vince Camuto's TWO labels name an *ikat* inspired textile for both a peasant blouse and a dress Prairie Tapestry. This was described by Lord and Taylor as *ikat,* Nordstrom calls it *ikat* inspired, and Bloomindales says it is inspired by authentic *ikat*. However, those terms were in the description which some consumers don't read; the main label often called it *ikat*. It is good that there is a more accurate

Fig. 5 Silk *ikat* shawl from author's collection purchased from Cloth Roads (2015). Photographer W. Hehemann

description offered for consumers who want more information before making a purchase. This is especially helpful for those shopping online who prize authenticity. Price should be an indicator, but sometimes consumers want to believe they are getting an authentic *ikat* for a bargain price! The other concern is how the textile was designed or was it simply digitized.

Rusuljon Mirzaahmedov, who started the Crafts Development Center in Margilan, is featured on a card from the International Folk Art Marketplace (Padilla 2004) used to help artisans tell their story. The success of the International Folk Art Marketplace has led to their web-site and online sales at www.ifamonline.org. This helps create more jobs and could improve quality and innovation. A web-site from Uzbek Craft, based in Portland, Oregon in the U.S., offers textiles on their web-site http://www.uzbek-craft.com/en/4-ikat-fabrics. Sometimes it is better for someone living in the country of marketing to select items for sale because they should have a better understanding of what their consumers want. Some companies with web-sites also have special events or even retail outlets to offer these special products. Cloth Roads had a trunk show around the December holidays in the U.S. Museum shops have also been selling more authentic textiles and other products.

Ikats are found in many countries including Mexico, Central America (Guatemala), South America (Argentina, Bolivia, Ecuador, and Peru), Madagascar, Yemen, New Zealand, Micronesia, and throughout much of Asia (Cambodia, Central Asia, China, India, Indonesia, Laos, Malaysia, Philippines, Taiwan, and Thailand).

Mary Connors author of *Lao Textiles and Traditions* Connors (1996) points out that the *ikat* technique was used as an early form of patterning on textiles. In Fig. 6, which Connors shot in Northern Laos, a Lao weaver proudly holds a weft *ikat* she produced. Some companies are giving more information and a photograph of the weaver; TOL in Bali does this. This connects the purchaser with the weaver. Some even state who made the yarns and who dyed the textile or yarns.

Over the last few years there has been a strong revival of interest in *ikats* for accessories such as scarves, purses, apparel, and interior textiles. This may relate to more interest in world culture, an increase in international travel, an increase in books on traditional textiles, ikats, and travel destinations, and the increase in authentic textiles offered for sale on the internet and in museum shops.

Textile designers often travel the world looking for motivation and/or new patterns. The author was in Tenganan Pegeringsingan when a designer from New York City came to visit. As the weavers opened up, the author was thinking "don't rip off their designs without giving them credit."

Some think it is more inspirational to visit the field for inspiration as Zandra Rhodes did for many of her painted silk textile collections used in several apparel lines. As design cycle times have shortened, only those truly looking for unique inspiration have the time for visits to far away production sites. With digitization, designers only have to find an image online, in an art book, or on a vintage textile to capture it.

Artists Gasali Adeyemo from Ghana, who now lives in Santé Fe, New Mexico, sells his work online at http://indigoarts.com/galleries/tie-dye-and-batik-textiles-west-africa. In addition, he teaches and attends conferences including the Textile Society of America 2014 to demonstrate and expose textile enthusiasts to Yoruba *adiri* batik. New Mexico even publishes a brochure called New Mexico Fiber Arts Trails: A Guide to Rural Fiber Arts Destinations (New Mexico Arts, n.d.).

Museum exhibitions also serve as inspiration; the increase in textile and clothing exhibitions by museums world-wide draws crowds and gives designers a close-up look at exquisite examples of techniques, colors, and uses. Some museums and universities such as Fashion Institute of Technology (FIT) in New York City have study collections for designers; their collection has a high quality old *geringsing* textile which has high quality depth of shade, a prized traditional characteristic.

Other designers settle for speed by trolling the internet, art books, and museum web-sites, where textiles are shown for inspiration. Some simply digitize and copy without giving any credit, or mash up the old with new interpretations thrown in, which can be jarring to people of traditional culture, although some involved in traditional cultures are also starting to experiment.

Fig. 6 Woman holds her
weft *ikat* weaving in N. Laos.
Photographer M. Connors

Today an increased number of consumers choose to express themselves by
acquiring and using unique textile products from scarves to apparel to interior
products such as throw pillows. Some express *ikat* fatigue on web-sites; however,
it is hoped that high quality traditional textiles continue to appeal to their key market segment, the authentic lovers of *ikats*.

2.1 Ikat Background and Analysis

The word *ikat* comes from the word *mengikat* which means to tie or to bind in
the Bahasa Malay/Bahasa Indonesia language, which was used along ancient sea
trade routes where many *ikats* were traded. The famed Indian silk *Patola* double
ikat textiles and other traditional textiles from India were traded in Indonesia,
Thailand, and beyond, where they frequently became used as status markers. *Ikat*
refers to both the technique and the textile. *Patola* cloth had a profound influence
on pattern and the *ikat* technique used in a wide region including in Southeast
Asia.

Ikat textiles literally tie or bind the yarn before dyeing and weaving just as tie-
dyed textiles are tied off in sections before dying, which allows both to resist the
color. The complexity added is that one must have a clear vision of the pattern

required while doing the tying. The dyed yarns can be two or more colors as shown in Fig. 1, which was tied evenly throughout the yarn bundle, then dyed blue.

Working with *ikat* yarns proves more difficult in general than tie-dying a textile because handling yarns is very difficult. The *ikat* yarns must be properly aligned to create the intended patterns; some movement to create the indeterminate line is expected, but too much can result in an error which would prevent the piece from being of the highest quality. Because *ikat* yarns generally appear to have indeterminate patterning sometimes described as hazy, they can forgive some errors, so exact precision of the hazy lines is not required by all *ikat* makers, especially those doing single *ikat*.

2.1.1 Classification of *Ikats*

The various types of *ikats* are classified as single, double, and compound. Single *ikats* use tie-dyed yarn in either the warp or filling direction. Warp *ikats* use *ikat* yarns in the warp or vertically aligned direction. This technique is easier to control because the weaver can adjust the yarn after it is on the backstrap loom to correct the pattern. Weft *ikat* textiles use *ikat* yarns in only the weft or filling direction. The weaver controls the patterning here by adjusting the weft yarn by hand or by using a pick to position the thread as seen in Tenganan Pergerinsingan, Bali. In most cases this is slower than a regular weft yarn insertion.

A true double *ikat* is created by tie-dying both the warp and filling to create color patterns at the weaving junctions of the two yarns. This is infinitely more complex than weaving a single *ikat* because the correct colors have to meet at the juncture points of the warp and weft yarns to make the pattern correct. This again may involve some maneuvering of the filling yarn to create the desired effect. Curvilinear patterns are more difficult to weave than linear ones.

Double *ikat* textiles are so difficult to make that historically they have been produced only in India, Indonesia, and Japan. Production in India is in Gujarat and in Japan on the island of Okinawa. In Indonesia, production is limited to Tenganan Pegeringsingan in Southeastern Bali. A description of changes in this village is presented in this chapter. Tenganan Pegeringsingan is the most accessible location to view the double *ikat* process.

In India, men are in charge of production and weaving whereas in Tenganan Pegeringsingan, Bali both the tying and dyeing are done by women with men sometimes assisting with dyeing. Men also make the loom and equipment.

Compound *ikats*, produced in Japan, have *ikat* yarns in both the warp and filling; however these yarns do not intersect to form a pattern. Hence they do not require the precision required in making the various colors intersect at the precise point that double *ikat* textile production requires.

After classifying the type of *ikat*, the researcher or observer usually analyzes the type of fiber used. Fiber influences the luster in the resulting textile, with silk being the most lustrous as in the *patola* cloths. Cotton, especially the very short staple type used in Tenganan, gives a dull or matte look. Cotton is used in Okinawa and other parts of Japan where complex *ikats* are made; silk is also used in *ikats* in Japan. Uzbekistan, Laos, Cambodia, and Thailand use silk.

Next, non-destructive examination of the yarns with a loupe may allow astute consumers to determine whether the yarn was hand or machine spun. Hand-spun yarns are generally more highly prized by collectors of traditional textiles, including those made more recently than machine-made-yarns. TOL has encouraged the use of hand-spun yarn. In Tenganan Pergerinsingan, the yarns are purchased on the island of Nusa Penida, off the main island of Bali. In some cases, yarns may be seen that are not on the loom, which offers a good way to understand more about the yarn.

Next, the dye process is analyzed. In some cases it is difficult to determine unless the dyeing for that specific textile was observed or the information was obtained from a reliable source. Demonstrations are just that, even with commissioned pieces. Collectors of antique textiles and some new textiles prefer natural dyes for the nuances and skill involved. For some buyers of new pieces, i.e., scarves, the buyer is more concerned with the general aesthetics. Some may also like the fact that they are helping weaving communities and or women artisans.

Some who are involved with traditional textiles such as Carol Cassidy prefer to use commercial dyes because it is easier to control the shade for several yarn dyeing batches. This is more likely to result in the same color throughout the textile for large quantities and repeat orders. Plants vary in their concentration or strength of active colorant. This is important if one is doing traditional textiles in larger quantities for draperies, upholstery, uniforms, or throw pillows. This is an aspect of color quality control.

Natural dyes often give more variation which is what most collectors of ancient traditional textiles prefer; this is similar with rug collectors who use the term *abrash* to denote an appearance of many colors in a textile or rug.

Natural dyeing often requires more skill and is considered both an art and a science. Sometimes the art and science of natural dying is lost or declines as was the case in the village of Tenganan Pegeringsingan where cotton yarns are first steeped for a number of days in *kemiri* nut oil with soda ash to help achieve deep reds; it appears tan or yellowish in the motif or lighter.

Next, the tied yarns are sent to the neighboring village of BugBug where the yarns are dyed using indigo because it is taboo to do indigo dying in the village. The weaver picks an auspicious day for dying. The indigo dyer must dip the yarn in indigo and air dry it several times to achieve a good depth of blue which is more highly prized both by both the weaver and by the person who purchases or uses the textile.

There are several reasons for the shift in color and depth of shade, but the decline in the quality of indigo dying by the woman dyer in BugBug led to a shift to a male dyer by the owner of one of the larger shops in the village of Tenganan Pegeringsingan. Females were traditionally those who knew about the plants for dyeing and other purposes in Southeast Asia. So this was a radical shift. They are then returned to the village for dyeing red with *Morinda citrifolia*.

In Jogjakarta, Indonesia in Central Java, Winotosastro Batiks had originally used indigo in batik production; however they switched to commercial dyes. In the late 1990s they started experimenting with re-introducing natural dyes. Ardianto Pranata in Jogjakarta also had a line of silk organza scarves made with natural dyes. Both labeled the products as having been dyed with natural dyes.

A test was deployed at a batik competition in Jogjakarta which required entrants to use natural dyes to determine whether true indigo was used. Recently, in looking at a new traditional piece from the researcher's collection, one color raised suspicion as to whether it really used all natural dyes. Although the researcher has seen the natural dyeing process in the village it does not mean that this textile did not use a synthetic dye to enhance one or more colors.

Some online web-sites simply do not state whether a natural dye was used because it would be difficult to monitor. A few web-sites do warn the consumer of color variations within the product—something that those who understand the process would understand and often prize the variation. This could indicate use of a natural dye. However, a new consumer might view it as an unacceptable flaw.

Ikat dying starts after the tying is completed. For simple two colored *ikats*, the dying is straightforward—tie the yarns, dye the yarns which may involve repeating the dipping and drying process, dry the yarns, then use the yarns in weaving. The yarns that are used in the warp may require special handling, i.e., gentle brushing to separate the yarns so they won't stick together in weaving. Deep reds and blues require frequent repeat dipping. Multiple colored yarns such as red over blue becomes more complex and requires retying. In Tenganan, overdyeing of red using *Morinda citrifolia* over indigo was historically used to give deep orange, burgundy, and browns.

2.2 Weaving Ikats

Many collectors and visitors to weaving sites want to understand the weaving process. Therefore, many sellers of traditional textiles in their village, workspace even if communal, or at small-scale cottage industry factories now see the benefit of welcoming tourists and explaining the process—which of course mean the possibility of increased sales.

In Nanjing, China, Jin Wen, a master weaver of *Yu Jin* embroidery initially found it difficult to sell even small framed samples to domestic tourists who came to see the famed draw loom weaving. Later this changed, and he opened several retail outlets which sold to both domestic and international tourists. Increased purchase by domestic tourists increased as China's economy expanded. He also designed products for special sports events in China.

Online web-sites such as Ock Pop Tok (OTP) http://ockpoptok.com/, Cloth Roads www.clothroads.com, and TOL www.threadsoflife.com use blogs, provide high quality photos, and give information about their products to educate the consumer. OPT has lodgings near their location and they offer courses for the interested textile enthusiast. TOL also has a strong presence on Trip Advisor, and you can walk online down their street, Jalan Kajang, to their shop. This helps tourists who have difficulty finding their small storefront as well as giving a realistic feeling before leaving home. They too have limited lodgings nearby. Many hotels in Indonesia, including the luxury Aman hotels, use traditional textiles in their décor.

The next step is weaving the *ikat* textiles; some are produced on the ancient backstrap loom as are textiles in Tenganan. The weaver must keep tension on the warp yarns by wrapping a strap around her back and extending her feet out flat often on a wooden portion of the loom. The width is often the width of the hips or slightly wider with experienced weavers; in Sumba weavers use a wooden attachment to widen the attachment so the female weaver can weave wider widths. In Tenganan Pegeringsingan a wider cloth is achieved by hand sewing two together to produce widths to drape and wear around the body; both males and females use these textiles for ceremonial purposes. The warp yarns remain uncut on certain ceremonial textiles.

Photographs and a description of the process of weaving Indian *patola* is described on Cloth Roads' Blog (Deatley 2015). Their warp yarns are longer than those used in other areas where double *ikats* are made; the weaving is done by men. Wooden upright floor looms are used in Bali to make cotton single weft *ikat, endek*.

2.3 Tying Ikat Yarns

In college, the author bound rubber bands on opaque blue opaque tights to create chocolate brown with blue irregular circles peeking through; only later did she learn that these are the classic colors used in batik in Central Java, Indonesia. Today, this still serves as a good entry point for students to study dying techniques. After observing both techniques, it is generally considered easier to do basic tie-dye on garments or flat textiles than on yarns. The same technique can be used to bleach out denim or other fabrics. Videos from UCLA Fowler Museum of weavers from Indonesia are also shown to students who usually think this is something they would never see. Hopefully they are surprised, as was the author who read Bühler's articles in *Ciba Geigy Review*, a publication by the dye company Ciba Geigy, never dreaming that she would visit these places. With globalization and virtual reality, more students have a chance to understand these places and textile techniques.

There are two broad methods used for tying yarns. One uses a special frame for wrapping yarns then binding them, and some put them directly on a frame. In Tenganan Pergerinsingan there is a different frame used for tying the warp and filling yarns (Bühler et al. 1975/1976). For tie-dying yarns some cultures use natural materials also illustrated in Bühler et al. (1975/1976); however, in Tenganan the use of polypropylene has replaced natural materials.

3 On the *Ikat* Interstate

After purchasing the *hinggi* in Los Angeles, the next stop on the *ikat* interstate was up to Los Angeles in 1972 to fall in love with traditional techniques at an exhibition of ancient Peruvian textiles with featured speaker Junius Bird. The breadth and difficulty of techniques, some of which were lost, intrigued the author, who considered herself more interested in textile technology and science up to this point. Interestingly, it was my graduate student who encouraged me to attend this historic event.

Next it was down to Mexico City in the early 1970s when I was finishing my courses for my doctoral degree and needed a break. There I saw traditional *ikat jaspe* shawls known as *rebozos*, other traditional textile techniques, and wonderful museums. Later I went to Panama to see Cunas wearing *molas*; then on to South America to see alpacas at Machu Picchu and to see traditional textiles in museums. Here I first saw women walking and using a drop spindle to spin alpaca. Wonderful alpaca products were available.

My first large collecting trip out of the country was to Guatemala in 1984 where I took all my books on Guatemalan textiles and two large custom Cordura™ bags to carry my bounty. I would purchase textiles by day and study and photograph them by night.

3.1 Asia, Indonesia, and Tenganan Pegeringsingan

The author's first trip to Asia was planned when working in the fiber industry on the export of branded premium fiber to Asia, which included working with many of the Asian fiber processors and companies producing or sourcing apparel there. It was evident that Asia was changing rapidly and that it was imperative to see it before it changed! One guy at work mentioned Bali was nice. With the help of a frequent flyer program, Beijing, Singapore, Indonesia—Jakarta, Jogjakarta, Solo, and Bali were next. The author found more resist textiles that she could resist.

After getting some education on Indonesian culture and art, including textiles, by attending most of the events held in the U.S. for the Year of Indonesia, including textile exhibitions at leading museums and lectures, the author was ready for Indonesia. When *Splendid Symbols* (Gittinger 1985) was released, the cover featured a photograph of young women from Tenganan Pegeringsingan swinging on the handmade wooden Ferris Wheel wearing their *geringsing* textiles. The author's goal was to see this ceremony.

Fortuitously, the author made it to Tenganan Pegeringsingan on her first trip to Asia in time to see the finale of their month-long *Usabah Sambah* ceremony, which included the swinging and the *mekaré-kare* also referred to as *geret pandan* or the famous *Pandanus* Leaf battle. This is considered to be the culmination of the month-long ceremony; it is the event that has always drawn the most tourists to the month-long event. Note that spelling and names vary greatly.

It was impossible to find out the dates of the ceremony unless you were on the ground in or near the village; it still remains very difficult to learn the dates for this event because the priest sets the dates using their calendar.

The author returned every year, sometimes many times a year, especially between 1991 and 1996 when she worked in the Republic of Singapore. From 1996 to 1999 the author lived in Indonesia where she had a research permit from The Indonesian Institute of Sciences known as Lembaga Ilmu Pengetahuan (LIPI) to conduct field work on Indonesian textiles, including those in Tenganan Pergerinsingan.

During this period, tourists hovered mainly in Kuta Beach; Picard (2003) referred to this as "tourists in quarantine" which means that officials were trying to contain tourists to the southern part of Bali near Kuta Beach, which is close to the airport, to prevent them from influencing the Balinese culture in other areas. At that time, transportation outside the quarantine area was difficult so the strategy worked for a time until more and more tourists started coming to Bali.

Transportation to the east side where Tenganan Pegeringsingan is located was difficult. Most tourists had to take a *bemo,* shared public transportation, which took several hours. Few private cars were available and there were no rental cars or private tour vans or buses operating in 1985.

Those wanting to go to Tenganan Pegeringsingan were mainly cultural tourists many of who came to see and experience the unique village, their unique *selandang* gamelan, and/or their renowned double ikat *geringsing* textiles. Collectors and scholars of traditional textiles knew about the *geringsing* textiles. The Museum fur Volkerkunde in Basel had purchased and exhibited some prime examples of complete ceremonial *geringsing* dress from villagers. Bühler and Ramseyer wrote about how *geringsing* was produced (Bühler et al. 1975/1975); Ramseyer made a video showing the complete process which takes a long time to complete because of the tying and lengthy dying. The author has also documented and photographed the entire textile process, changes, and ceremonies, including those performed by unmarried women during the month-long ceremonies of *Usabah Sambah.*

Estimates of the amount of time needed for completion range widely because women do not work 8 h days. This allows them more freedom to attend to their families and home when earning money. Also, simplification of the dyeing process as well as more weavers making smaller textiles, using simpler motifs, does not allow an accurate estimate of time to be made by the author.

The purchase of outstanding museum quality complete dress allowed some of the sons of outstanding weavers to open their own businesses in the village and one, *Geringsing* Homestay, opened in Candi Dasa, the town where most tourist visiting Tenganan Pegeringsingan stay, because outsiders are not allowed to stay in the village overnight. Today, package tours stop at several sites in Eastern Bali and they may just allow tourists to spend a limited time in the village. There are now high end hotels nearby.

In 1985, the first year the author visited, three women, purported to be in their 90s, were still weaving with their legs flat on the floor when weaving on their backstrap looms. The next year, these women were no longer weaving. At this time, there were a few middle aged weavers and one younger weaver. Weaving was in decline and many thought it could never be revived. Most interested in high quality weaving were not interested in the outcome. The author was interested in following what happened, which took an unexpected turn when increased tourism helped to revive the weaving and *ikatting* culture of the village.

4 Tenganan Pegeringsingan

4.1 Village Background

Previously, the walled village of Tenganan Pegeringsingan had been closed to all outsiders, including other Indonesians and Balinese, as verified by Pia Alisjabana, founder and CEO of Indonesia's fashion publishing empire The Feminina Group. Mary Kefgen, a former professor from California State University at Long Beach, visited the village and said that a couple donned their ceremonial *geringsing* textile dress for her to photograph. There were, of course, no tourists other than herself.

Geringsing textiles were well known on the island and also very valuable because they were considered powerful and believed to cure illness. *Geringsing* means illness averting. Some cloths have a small swatch cut out of them; the cloth may have been dipped in boiling water and the tea drunk or the swatch used as an amulet. In 1998, the author attending a tooth filing ceremony where three cousins had the ceremony performed; they used a small *geringsing* textile to help them through the process of filing teeth with a regular metal hand-held file.

Bali has a majority of Hindus in the population, even with the influx of Muslims from Java to work on roads, construction, and restaurants in Bali. Most Balinese Hindus follow the caste system; however, in Tenganan Pegeringsingan they do not. Nor do they cremate their dead as do other parts of Hindu Bali. However, they do place a *geringsing* cloth over the genitals of the deceased. A form of recycling exists when they sell these textiles to unsuspecting tourists.

Some refer to the people of Tenganan Pegeringsingan as Bali *Aga*, Bali *Mula,* or original Balinese (Hobart et al. 2001); however, the people of Tenganan Pegeringsingan and others do not use these terms. Their history is shrouded in mystery, but one thing is certain—the village puts prime emphasis on keeping the village pure. It has a large number of ceremonial days in which qualified villagers participate by helping with preparing and performing ceremonies which includes wearing of proper ritual dress.

The village, situated in Southeast Bali, has about 1500 hectares of prime land (Ramseyer and Berger 2009) which share croppers help to farm; hence they were considered fairly well off compared with the average Balinese. Most villagers do not work outside the village. However, working outside the village has become slightly more common. Generally it is the males who work outside the village.

However, in the summer of 2014, a female was working at a restaurant at a small hotel in Candi Dasa, but she was still able to live in the village.

As weaving increased, the author asked women whether they learned to weave as young girls. Those who had previously learned to weave in one of three unmarried girls clubs in the village were asked why they returned to weaving; most cited to get money to help children; they had to buy school uniforms—and some said candy. There are still no consumer shops in the village, but outside the village small shops sold individual size packets of shampoo initially. Now one might see larger bottles of shampoo in a home. Money became a necessity. The village also had one pay phone—now many have cell phones.

The fame of *geringsing* cloth is well documented on the island. The author visited several families in Bali outside Tenganan who would insist upon bringing out their *geringsing* textiles. While researching *songket* weaving in Sideman, the family insisted on showing their small *geringsing* which had been used in tooth filing ceremonies to protect the participant. Many families still believe in the protective power of *geringsing* textiles.

4.2 Ritual Dress and Sacred Ceremonies

The villagers, especially the young men and women in the associations and their advisors, wear prescribed clothing, including that made from draping *geringsing* textiles, during their sacred rituals and ceremonies, each having specific clothing. In Fig. 7, young unmarried women walk down the ancient streets of the village to a ceremony in their complete *geringsing* dress. The young women's dress consists of a large uncut cloth or hip wrapper composed of two large *geringsing* textiles hand sewn together, then draped and wrapped on the body, a breast wrapper, and the extended frontal piece (Ramseyer and Berger 2009). The frontal view seen in Fig. 8, shows the red plaid hand-woven underskirts.

The women are standing on their *balé* (pronounced ballet) as they prepare for ceremonies later that day during the month-long *Usaba Sambah*. A range of *geringsing* patterns and motifs are visible with the young woman on the far right having a high quality deep dyed textile with an intricate and difficult-to-make motif. The woman below is the group's advisor, who is also wearing prescribed dress and helps the unmarried women execute their ceremonies correctly which helps to keep the village pure. The young men and women both have three associations.

Young men of the village chop and prepare bundles of *pandanus* leaves for the fight the next day in front of the *Balé Agung* in Fig. 9 by cutting the long leaves or stems into smaller pieces and then wrapping them into small bundles for each male to use when he goes into battle.

The men in Fig. 9 are dressed casually in either printed batik or *endek* weft *ikat*; the workers in front wear an overlaid white checked hand-woven textile. When men are on the *Balé Agung*, they must not cover their chests and they must wear a hip wrapper. This is the area where the village males watch the fight.

Fig. 7 Unmarried women walk the streets of Tenganan Pegeringsingan in their ceremonial *geringsing* dress during *Usabah Sambah*. Photographer K. Crippen

Fig. 8 Unmarried women perform ceremonies in ceremonial *geringsing* dress on their *balé* in Tenganan Pegeringsingan. Photographer K. Crippen

Previously invited guests, such as researchers, or those known to the village, such as a doctor who had worked in the village, are invited to the *Balé Agung* to watch the activities and interact with the village men.

Men who are injured from the *pandanus* thorns have a turmeric-based potion applied. Another potion, *arak*, rice wine, also flows to the men and guests. Therefore it is not surprising that many tourists all want to get a prime location

Fig. 9 Unmarried men work at their *Balé Agung* preparing bundles of *pandanus* leaves for the *Mekaré-kare* which is held on a stage in front of their *balé*. Photographer K. Crippen

out of the sun in front of the stage to view the battle. Interestingly, once tourists see other non-Balinese on the stage, they too often think they want to get up there. However, they do not understand that proper dress is required. Women are required to not have anything covering their shoulders. Some women have on a *kebayah*, a lace jacket, which is the national dress for women in Indonesia, along with a hip wrapper. They think that is proper dress. Recently, researchers and others have no longer been invited to participate on the *Balé Agung*. The unmarried women watch from their *balé*. No guests are allowed to stand on their *balé*.

On the day of the *mekare-kare*, the male musicians of the gamelan *selandang*, villagers, trance dancers, and tourists circumambulate the village with the musicians playing their music. Then the fight starts on a makeshift stage with men from the village fighting one another briefly. Here the two men standing on the left in Fig. 10 are serving as referees. The young unmarried man on the far left is dressed in a high quality traditional *geringsing* hip wrapper. The referee on the right in white is wearing a hip wrapper *dhoti* style to allow for maximum movement when he moves to break up fighting if it becomes too rowdy.

Textile enthusiasts including the author are often concerned that such high quality textiles are damaged during battle or become soiled with perspiration which is visible on one man's back. These textiles might be hung out to air, but most would not be hand washed. On Saraswati Day, the villagers place communal textiles from the store house in the sun on mats on the ground to air.

In 1985, more men were wearing *geringsing* when they participated in the mock battle or *mekare-kare*. However, as time went on the villagers started wearing less expensive hip wrappers, either real batik *tulis*—hand waxed, batik *cap*—hand stamped, printed imitation batik, or single weft *ikat, endek* produced on

Fig. 10 Referees for the *Mekaré-kare* with the *one on the left* in ceremonial *geringsing* dress and the *one at the right* in a *white dhoti*. Photographer K. Crippen

an upright floor loom in a nearby village as is worn by the man on the far right. The young man in the front adjusts a real waist wrapper of *geringsing* cloth. One informant in 2014 mentioned that he used to wear his old *geringsing* but took it off when it was his turn to fight. Some have sold their old *geringsing* textiles. It is more important for the female to wear these textiles. Older high quality *geringsing* textiles serve as status markers.

Men use their hand-made basketry shields which were originally made in the village as shown in Fig. 11. Here the bundle of *pandanus* leaves used in fighting are visible. One man in the village opened a shop in his home which sold plaited baskets and purses; today this is the major export item from the village. Villagers contract workers from the nearby hills to do the work.

After the *Mekaré-kare*, a communal meal which has been cooked by the men is offered to the unmarried participants. Families also have feasts complete with *babi guleng*, whole roast pig. Women make offerings and later take them to the temple, carrying them on top of their heads.

Other temple activities includes the unmarried women performing the *rejang* dance. In Fig. 12 a young girl await the start of the dance performance in the main lower portion of the open air temple and Mary Connors is wearing a lace *kebayah* and hip wrapper, which is the correct temple dress for women tourists, as she took photos of the young women after asking permission. Both men and women must wear a hip wrapper.

Tourists visiting during this time may also visit shops inside homes. New *geringsing* cloth was sold exclusively inside the village until recently. In 2014, the author saw a new *geringsing* for sale outside the village for the first time at TOL. Some village families continue to sell high quality older textiles, which are also sold by dealers and traders throughout the world. During these ceremonies,

Fig. 11 Unmarried men fighting during the *mekare-kare*. Photographer K. Crippen

Fig. 12 Mary Connors in lace *kebayah*, considered the national dress of Indonesia, photographs a young women getting ready to perform the *rejang* dance inside the Hindu village temple in Tenganan Pegeringsingan during *Usambah Sambah*. Photographer K. Crippen

shoppers may get a better deal on antique and new textiles because the families need money to pay for their feasts. Downward valuation of the Indonesian *rupiah* also sends buyers to the village.

In 1985, the first year the author visited the village, she found four shops in homes where women were weaving and offered demonstrations. This allowed the visitor to understand the complex process even better. The women would demonstrate the weaving process in the hope of making a sale.

However to see the yarn tying process and dying, one had to return frequently. Figure 13 shows filling yarns tied ready for dying. These yarns have already been stained using *kemiri* nut oil from the candlenut—they sit in the oil for 42 days. This helps to deepen the shade of red.

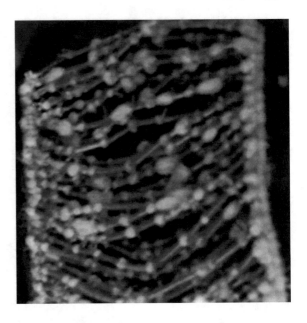

Because the *geringsing* textiles have a unique aesthetic which is somewhat an
acquired taste, shop owners started to sell other textiles; they also sold T-shirts and
endek which might even emulate a *geringsing* cloth. Dealers from other islands
came to sell the shop owners textiles from other islands which initially sold better
than *geringsing* textiles because they were less specialized aesthetically. However,
over time, more tourists who were interested in textiles and the village started to
purchase *geringsing* textiles.

The smaller widths are the most widely sold because tourists can more readily
afford these. Three different motifs of the small size textile are shown in Fig. 14.
The villagers just weave their traditional motifs. The villagers do not make new
patterns but continue with traditional patterns. Breast wrappers used in ceremonies
do not have their warps cut. Most textiles sold to tourists do have their warps cut
and are made specifically for the tourist trade. The light colored selvage is visible
on the edges of the textiles. The villagers do not fold the selvage under or hem
the edges of the cloth to make the selvage invisible. The colors achieved in 2014
appear to be improving in some cases; however, the key improvement needed continues to be an increasing greater depth of shade.

A villager weaves on her backstrap loom in a home in Fig. 15; note her legs
are stretched out to maintain tension on the warp yarns on the loom. In this household, there were many women weaving during the daytime and one was tying.
The weaver's T-shirt from The Landmark Hotel in Las Vegas was either traded or
donated to her. Photographs of women weaving when wearing a *geringsing* breast
wrapper are staged; women weave in whatever they are wearing. After they fin-
ish weaving for the day, they roll their weaving up and cover it to store it. One
woman weaving a wide textile advised I. Komang Sumanta from the village, who

Fig. 14 Three different *geringsing* motifs in small size textiles in Tenganan Pegeringsingan. Photographer K. Crippen

was with me that day, that she had to go to feed her pigs. This is the beauty of having traditional weaving in a village setting in that it allows women to adjust their schedule to take care of all their chores and their families.

In Sideman, a woman covered her *songke*t weaving at lunch time—her babies' lunchtime. Women who have to go to factories to work must live in dormitories and do not get to participate in family life. Hence, as the demand for traditional textiles has increased, more families are able to make a living by making traditional textiles, many in or near their village.

The key event for the author of the month-long *Usabah Sambah* was the young women wearing their traditional *geringsing* dress riding the hand-made wooden Ferris wheel hand-powered by young men, shown in Fig. 16. In 2014, a double seated Ferris Wheel was seen.

Double *ikat*, one of the world's most difficult textiles to make, requires the most skill in tying off the yarns correctly before dyeing; the pattern is visible only after weaving. Single *ikats* only require tying yarn in one direction, the warp or filling. Double *ikats* not only require tying of both the warp and filling yarns, but also getting them to align properly during weaving to create the desired pattern, which is inherently difficult.

The village has gone from being closed to outsiders to being described as a living museum—a term that many villagers and some tourists find limiting. In *The theatre of the universe: Ritual and art in Tenganan Pegeringsingan*, Ramseyer and Berger (2009) also documents changes in the village.

Initially, the young women learned techniques associated with the entire process in their girl's association, but that changed as they boarded in other villages in order to attend school. This had a major impact on the decline of weaving and new

Fig. 15 Woman weaves small *sanan empeg* motif on backstrap loom in Tenganan Pegeringsingan. Photographer K. Crippen

women learning all the steps in the textile process. There are three young men and young women or girl's associations. Each has a sponsor. It was in their sponsor's home that the young women learned the textile arts. As young women left the village to attend school and board outside the village, interest and the ability to study the textile arts started to decline.

However, the eventual increase in women weaving after 1985 was driven mainly by tourism and the villagers' need for cash. Some of the women returning to weaving had learned how to weave, but had stopped. They didn't have a market for their products because there were few tourists; the tourists who came wanted high quality weavings or antique heirloom *geringsing* textiles. Some women were not outstanding weavers, but money offered the incentive they needed to improve.

One key concern was not how many women were weaving but how many younger women were starting to weave at a higher level, making more intricate and wider patterns. The depth of shade of the textile remains a key concern.

Tourism was initially mainly limited to cultural tourists. Villagers slowly started opening their homes for tourists to see the weaving and sometimes they

Fig. 16 Young unmarried
women in *geringsing* dress
swing on the Ferris Wheel
in Tenganan Pegeringsingan.
Photographer K. Crippen

got to see the tying of yarns or dying. The weaving is actually the easiest step
and tying is the most complex. As tourism increased slightly, a few homes offered
sodas and snacks for purchase.

As tourism increased and transportation became easier, tourists started arriving
in cars, vans, and buses. Many shops increased their inventory but mainly of textiles from other islands, T-shirts, and single *endek ikats*, some imitating the double *ikats*. Tourists wanted to see the process, but initially they did not always buy
a *geringsing* textile. *Geringsing* has a very specialized aesthetic. However, more
tourists who come now buy an authentic *geringsing* in the village.

The villagers are now also outsourcing the weaving. In the late 1990s two
younger weavers/dyers took a student from the neighboring village of Pandai
Sikit. This village parallels Tenganan Pegeringsingan; those who broke the marriage rules were exiled. Others exiled set up stores near the parking lot.

In 1985, most women were making the *sanan empeg* motif with linear
lines, which was easier to make than the more elaborate floral and *wayang* motifs
with curvilinear ones. Some women have started weaving wider textiles and more
patterns. Today, a few women are weaving wider widths.

High quality antique textiles are still retained by some families for ceremonial usage. The village market for new *geringsing* textiles is driven both by tourists and by unmarried women's participation in-ceremonies. Although young men initially wore *geringsing* for the month-long ceremony, now they often wear batik or single *ikats*.

Women weavers or those that sell textiles out of their home have contributed financially to their family in a village with few other businesses. The villagers are considered land rich and participate in a share cropping scheme; hence they are considered relatively affluent. Weaving is important work because the textiles help with the sanctity and purity of the village. The weavers help to sustain the village.

Reflecting back on the first visit to Tenganan during the *Mekaré-kare* and comparing it with the latest visit during this time in 2014, the village has changed immensely. Many interested in the continuation of *geringsing* weaving in Tenganan Pegeringsingan feared that it might disappear altogether or not make any progress. In 2014, the village is vibrant and *geringsing* weaving continues through the villagers and nearby residents' efforts. The villagers have made all of their own decisions in how to proceed without the help of development experts.

As the author ponders her time spent in the village, starting in 1985 and as a researcher from 1996 to 1999 in Indonesia, she realizes that she has seen a very special transformation of a village that has continued to weave and use *geringsing* textiles in their ancient sacred ceremonies and life cycle rituals. As shown in Fig. 17, the author always wears this single *ikat* during *Usabah Sambah* purchased in the village and purportedly made by people in the hills near Tenganan.

A high quality borrowed older *geringsing* is worn as a *selandang*, shoulder scarf, commonly worn in Indonesia. The *selandang* is removed as she proceeds to the *balé* to watch the *mekare-kare*. She is joyful that the weaving tradition and ceremonies continue and help with the sustainability of the village.

Fig. 17 Author in *geringsing* in Tenganan Pegeringsingan. Photographer I. Komang Sumanta

As the author examines the textile in front of her in Fig. 18 from Sumba, purchased at TOL, she is pleased that more tourists including textile collectors are now starting to purchase new Indonesian textiles which helps to sustain many weavers and their communities. At TOL, the author interviewed a woman purchasing a large Indonesian hand-woven textile who said she and her husband had an antique one, but they wanted this one to hang and they wanted to support the weavers. TOL has just been able to start shipping textiles by air mail from Indonesia which should help with making the global marketplace more accessible.

In Fig. 18, the author wears an authentic single warp *ikat* jacket from Berek which was purchased as ready-to-wear. She wears a small new *geringsing* as a *selandang* purchased in 2014 from Tenganan Pegeringsingan. Once when wearing this jacket, another textile researcher said, "You're wearing real *ikat*"! This was not a comment heard before when wearing this jacket because most people are unable to discern this, especially with just a quick glance.

This understanding of what is authentic is critical for weavers and artisans to maintain the demand for their hand-made traditional products. The benefit to weavers and artisans extends to their communities as many such as Judy Frater

Fig. 18 Author in weft *ikat* jacket wearing *geringsing* *selandang* and examining new warp *ikat* from Sumba purchased at TOL 2014. Photographer W. Hehemann

of Somaiya Kala Vidya and William and Jean Ingram of TOL and others have learned.

This global marketplace is also aided by the increased number of businesses specializing in selling these products on the internet, at craft fairs and events, or large international marketplaces, as well as small retail outlets in the country of origin or boutiques in other countries.

Authentic *ikat* textiles have been knocked-off, copied, digitized, and used as a source of inspiration in apparel, accessories, and interior textiles without acknowledgement of their origins. Some online comments express comments about seeing too many brightly colored *ikat* patterns—as if we are tired of this; let's move on to the next trend.

Whenever the fashion trend for *ikats*, imitation *ikats*, and those produced by digitizing declines, it is hoped that the market segment that cherishes authentic *ikats* can remain. In an interview, Judy Frater discussed the role of innovation in dealing with the artisans in India at Somaiya Kala Vidya, an institution which she founded and where she is currently Director.

In Tenganan, the weavers are using the same motifs and using the same traditional colors, although some of the deep old overdyed colors are no longer produced. It is probably good for the continuation of the authentic designs that not much experimentation has occurred.

The astute wearer, owner, and observer may feel a difference when they experience the authentic textile. The author certainly feels more joy wearing hand-tied *ikat* products than when wearing *ikatish* printed and perhaps digitized patterns.

Traditional ethnic textiles make the world a more interesting place and a respite from the uniformity of the mass produced textiles of fast fashion. The richness of watching the continuation of traditional textile production in person and the resulting use of apparel and other textile products used in daily, ritual, or ceremonial aspects of life reminds us of the beauty and importance of traditional ethnic textiles.

5 Conclusions

The global marketplace for traditional textiles including hand-made *ikats* continues to expand as more are exposed to global cultures and aesthetics through travel or events which feature unique products. The advantage for socially conscious consumers of supporting these types of products has increased. The internet has greatly aided the process where entrepreneurs can tap into the needs of global or regional consumers to offer edited collections. Previously, it was often difficult for a small-scale entrepreneur without experience in what international consumers prefer. Today, there are more people participating and interacting to bring more authentic products to the global marketplace.

Entrepreneurs who have been successful hopefully continue to help others in their villages to become successful as well. Increased travel for artisans to special events gives them an opportunity to interact with buyers and experience first-hand reactions of consumers, i.e., this scarf really is too big for most consumers or they love the color combinations, patterns, and style of the hand-made purse.

One major concern is the ease with which some traditional patterns can be digitized and knocked-off with no mention of the culture of origin. Patterns or garments may have names that do not reflect anything from the original culture.

As more consumers become educated about traditional textile techniques, it is hoped that more of them can trade up to consuming more authentic products instead of wearing just the knock-offs. The *ikat* image has been around for some time in the marketplace and some interior blogs mention boredom with the look. This means that producers of authentic traditional textiles must strive to understand their consumers. It is hoped that technology can help to unite the artisan or producer even further with the consumer and that more families can be allowed to tap into the global marketplace.

In addition, it is hoped that traditional textile usage in sacred ceremonies as described in Tenganan Pergerinsingan can gain more widespread understanding and appreciation by those attending or those interested in the village. Traditional textiles are used in both sacred and secular ceremonies, e.g., Chinese New Year; both often have regulations or guidelines surrounding usage which many are eager to understand. Video blogs and other technology platforms can allow more people to obtain additional understanding even if they can't go in person. This allows people to determine where they would like to visit.

The worst fear is that traditional textiles used in traditional dress continue to be lost to modern western style clothing usage. Consumer purchases and involvement can help the artisan communities stay alive. It can also help them to see that others value and appreciate their work. Continuation of textile traditions can help sustain communities spiritually and physically. Also, from a sustainability viewpoint, traditional textiles do not go out of style and hence they can be worn for much longer than the latest fast fashion.

Acknowledgements Lembaga Ilmu Pengetahuan (LIPI) or Indonesian Institute of Sciences for the research permit from 1996 to 1999

Puspita Wibisono, sponsor, and Director of The Textile Museum in Jakarta, Indonesia

I. Komang Sumanta—originally from Tenganan Pergerinsingan

The weavers and villagers from Tenganan Pergerinsingan

Patricia Mulready—co-author on articles on Tenganan Pergerinsingan

Mary Kefgen—for the loan of all her books on batik and traditional textiles

Evans Allen funding for research on Sustainability of bio-materials and communities—University of Arkansas at Pine Bluff—UAPB

Will Hehemann—Photographer and Communications Specialist—UAPB School of Agriculture, Fisheries, and Human Sciences

Brad Mayhugh—Photographer and Communication Specialist—UAPB School of Agriculture, Fisheries, and Human Sciences

Jameka Cottery—Stylist for photo shoot—UAPB student in Merchandising, Textiles, & Design Program

Flavelia Stigger—Research Technologist for sample preparation for photography and formatting the references—UAPB Human Sciences Department.

References

Adams MK (1969) System and meaning in East Sumba textile design: a study in traditional Indonesian art. Yale University, New Haven

Anon (2015) Gifts representing China's ethnic diversity and silk road fabrics elevate D.C. Museum to Top Collector of Central Asian and Chinese Minority Textiles. https://museum.gwu.edu/gifts-representing-china%E2%80%99s-ethnic-diversity-and-silk-road-fabrics-elevate-dc-museum-top-collector. Accessed 05 Dec 2015

Barnes R, Kahlenberg M (eds) (2010) Five centuries of Indonesian textiles: the Mary Hunt Kahlenberg collection. Delmonico Books-Prestel, New York

Bebali K (2011) Sacred cloth: meaning and usage in Balinese Hindu rituals. Indonesian Heritage Society, Jakarta

Bühler A (1959) *Patola* influences in Southeast Asia. J Indian Text Hist 4:4–46

Bühler A, Ramseyer U et al (1975/1976) *Patola und geringsing: Zeremonial tucheraaus Indien und Indonesien.* Austellungsfuhrer Museum fur Volkerkunde, Basel

Cloth Roads Blog (2015) https://www.clothroads.com/bandhani-artisan-expands-resist-dyeing-tradition-aziz-khatri/?v=7516fd43adaa. Accessed 03 Dec 2015

Connors M (1996) Lao textiles and traditions. Oxford University Press, Kuala Lumpur

Crippen LK (1994) Continuation and change in *geringsing* double *ikat* weaving in Tenganan, Bali. In: Proceedings, Indonesian and other Asian textiles: a common heritage. Indonesian Heritage Society, Jakarta

Crippen LK (2015) Continued Change in *Geringsing* weaving in Tenganan, Bali. In: Proceedings of ITAA international annual meeting 2015. November 9–13. Iowa State University, University Digital Library, Ames

Crippen LK, Mulready PM (1995) Textile traditions and quality-of-life concerns in Southeast Asia. In: Sirgy MJ (ed) Proceedings developments in quality-of-life studies and marketing. Williamsburg, VA, pp 100–107

Crippen LK, Mulready PM (2001) Continuation and change in Tenganan Pegeringsingan Bali. In: Arthur L (ed) Undressing religion. Berg Books, London, pp 183–199

Crippen LK, Mulready PM (2012) Proper dress required: protecting the sacredness of the Tenganan Pergersingan, Bali *Usaba Sambah* ceremony utilizing dress rules. In: Proceedings of the international tourism studies association biennial conference

Crippen L, Mulready PM (2014) Spinning yarns—weaving textiles: creating authentic internet stories. In: Proceedings GAMMA global marketing conference at Singapore, July 15–18, Republic of Singapore, 2014

Deatley S (2015) Patan *Patola*: India's heirloom double *ikats*. https://www.clothroads.com/patan-patola-indias-heirloom-double-ikats/?v=7516fd43adaa. Accessed 03 Dec 2015

Eicher JB, Evenson SL et al (2015) The visible self: global perspectives on dress, culture, and society. Fairchild, New York

Gibbon KF, Hale A (1997) *Ikat*: splendid silks of Central Asia: the Guido Goldman collection. Laurence King Publishing, London

Gittinger M (1985) Splendid symbols: textiles and traditions. Oxford University Press, London

Hamilton RW (1994) Gift of the cotton maiden: textiles of Flores and the Solor Islands. Fowler Museum of Cultural History at UCLA, Los Angeles

Hamilton R (2012) Weavers stories from Island Southeast Asia. In: Fowler Museum at UCLA, Los Angeles. (CD included)

Hamilton R (2014) Textiles of Timor: Island in the woven sea. In: Fowler Museum at UCLA, Los Angeles

Hobart AR et al (2001) The people of Bali. Blackwell, Oxford

Ludington S (2011) Central Asian style. https://www.clothroads.com/what-is-ikat-central-asian-style/?v=7516fd43adaa. Accessed 03 Dec 2015

Ludington S (2015) Uzbekistan *ikats* then and now. https://www.clothroads.com/uzbekistan-ikats-then-and-now/?v=7516fd43adaa. Accessed 03 Dec 2015

Maxwell R (2013) Textiles of Southeast Asia: Tradition, trade, and transformation. Tuttle, North Clarendon

Murphy M (2013) *Bandhani* artisan expands resist-dyeing tradition: Aziz Khatri. https://www.clothroads.com/bandhani-artisan-expands-resist-dyeing-tradition-aziz-khatri/?v=7516fd43adaa. Accessed 03 Dec 2015

New Mexico Arts (n.d.) New Mexico fiber arts trails: a guide to rural fiber arts destinations

Ock Pop Tok (2015) http://ockpoptok.com/. Accessed 05 Dec 2015

Padilla C (2004) The work of art: Folk artists in the 21st century. In: IFM Media, Sante Fe

Peck A (ed) (2013) Interwoven globe: The worldwide textile trade, 1500–1800. Yale University, New Haven

Picard M (2003) Touristifcation and Balinization in a time of *reformasi*. Indonesia and Malay World 31(89):108–118

Ramseyer U (1984) Clothing ritual and society in Tenganan Pegeringsingan. Verhandlungen der Naturforschenden Gesellschaft, Basel

Ramseyer U, Berger U (2009) The theatre of the universe: Ritual and art in Tenganan Pegeringsingan. Museum der Kulturen, Basel

Ramseyer U, Hinz H (1977) The art and culture of Bali. Oxford University Press, Oxford

TOL (2015) www.threadsoflife.com. Accessed 05 Dec 2015

Printed in the United States
By Bookmasters